Chemistry in the Marine Environment

900428972 8

KV-486-141

WITHDRAWN
FROM
UNIVERSITY OF PLYMOUTH
LIBRARY SERVICES

Charles Seale-Hayne Library
University of Plymouth
(01752) 588 588
LibraryandITenquiries@plymouth.ac.uk

ISSUES IN ENVIRONMENTAL SCIENCE AND TECHNOLOGY

EDITORS:

R.E. Hester, University of York, UK
R.M. Harrison, University of Birmingham, UK

EDITORIAL ADVISORY BOARD:

Sir Geoffrey Allen, Executive Advisor to Kobe Steel Ltd, **A.K. Barbour,** Specialist in Environmental Science and Regulation, UK, **N.A. Burdett,** Eastern Generation Ltd, UK, **J. Cairns, Jr.,** Virginia Polytechnic Institute and State University, USA, **P.A. Chave,** Water Pollution Consultant, UK, **P. Crutzen,** Max-Planck-Institut für Chemie, Germany, **S. J. de Mora,** International Atomic Energy Agency, Monaco, **P. Doyle,** Zeneca Group PLC, UK, **M.J. Gittins,** Consultant, UK, **J.E. Harries,** Imperial College of Science, Technology and Medicine, London, UK, **P.K. Hopke,** Clarkson University, USA, **Sir John Houghton,** Meteorological Office, UK, **N.J. King,** Environmental Consultant, UK, **J. Lester**, Imperial College of Science, Technology and Medicine, UK, **S. Matsui,** Kyoto University, Japan, **D.H. Slater,** Oxera Environmental Ltd, UK, **T.G. Spiro,** Princeton University, USA, **D. Taylor,** Zeneca Group PLC, UK, **T. L. Theis,** Clarkson University, USA, **Sir Frederick Warner,** Consultant, UK.

TITLES IN THE SERIES:

1 Mining and its Environmental Impact
2 Waste Incineration and the Environment
3 Waste Treatment and Disposal
4 Volatile Organic Compounds in the Atmosphere
5 Agricultural Chemicals and the Environment
6 Chlorinated Organic Micropollutants
7 Contaminated Land and its Reclamation
8 Air Quality Management
9 Risk Assessment and Risk Management
10 Air Pollution and Health
11 Environmental Impact of Power Generation
12 Endocrine Disrupting Chemicals
13 Chemistry in the Marine Environment

FORTHCOMING:

14 Causes and Environmental Implications of Increased UVB Radiation

How to obtain future titles on publication

A subscription is available for this series. This will bring delivery of each new volume immediately upon publication and also provide you with online access to each title via the Internet. For further information visit www.rsc.org/issues or write to:

Sales and Customer Care Department
Royal Society of Chemistry
Thomas Graham House
Science Park
Milton Road
Cambridge CB4 0WF, UK

Telephone: +44 (0) 1223 420066
Fax: +44 (0) 1223 423429
Email: sales@rsc.org

ISSUES IN ENVIRONMENTAL SCIENCE AND TECHNOLOGY

EDITORS: R. E. HESTER AND R. M. HARRISON

13

Chemistry in the Marine Environment

RS•C

ROYAL SOCIETY OF CHEMISTRY

UNIVERSITY OF PLYMOUTH
Item No. 900 4289728
Date 21 JUN 2000 S
Class No. 551.4601 CHE
Contl. No.
LIBRARY SERVICES

ISBN 0-85404-260-1
ISSN 1350-7583

A catalogue record for this book is available from the British Library

© The Royal Society of Chemistry 2000

All rights reserved
Apart from any fair dealing for the purposes of research or private study, or criticism or review as permitted under the terms of the UK Copyright, Designs and Patents Act, 1988, this publication may not be reproduced, stored or transmitted, in any form or by any means, without the prior permission in writing of The Royal Society of Chemistry, or in the case of reprographic reproduction only in accordance with the terms of the licences issued by the Copyright Licensing Agency in the UK, or in accordance with the terms of the licences issued by the appropriate Reproduction Rights Organization outside the UK. Enquiries concerning reproduction outside the terms stated here should be sent to The Royal Society of Chemistry at the address printed on this page.

Published by The Royal Society of Chemistry, Thomas Graham House,
Science Park, Milton Road, Cambridge CB4 0WF, UK
For further information see our web site at www.rsc.org

Typeset in Great Britain by Vision Typesetting, Manchester
Printed and bound by Redwood Books Ltd., Trowbridge, Wiltshire

Preface

The oceans cover over 70% of our planet's surface. Their physical, chemical and biological properties form the basis of the essential controls that facilitate life on Earth. Current concerns such as global climate change, pollution of the world's oceans, declining fish stocks, and the recovery of inorganic and organic chemicals and pharmaceuticals from the oceans call for greater knowledge of this complex medium. This volume brings together a number of experts in marine science and technology to provide a wide-ranging examination of some issues of major environmental impact.

The first article, by William Miller of the Department of Oceanography at Dalhousie University in Nova Scotia, provides an introduction to the topic and an overview of some of the key aspects and issues. Chemical oceanographic processes are controlled by three principal concepts: the high ionic strength of seawater, the presence of a complex mixture of organic compounds, and the sheer size of the oceans. The organic chemistry of the oceans, for example, although involving very low concentrations, influences the distribution of other trace compounds and impacts on climate control via feedback mechanisms involving primary production and gas exchange with the atmosphere. The great depth and expanse of the oceans involve spatial gradients and the establishment of distinctive zones wherein a diversity of marine organisms are sensitive to remarkably small changes in their chemical surroundings. The impact of human activities on marine biodiversity is of growing concern.

The second article, by Grant Bigg of the School of Environmental Sciences at the University of East Anglia, is concerned with interactions and exchanges that occur between ocean and atmosphere and which exert major influences on climate. Through carbonate chemistry the deep ocean is a major reservoir in the global carbon cycle and can act as a long-term buffer to atmospheric CO_2 while the surface ocean can act as either a source or sink for atmospheric carbon, with biological processes tending to amplify the latter role. CO_2 is, of course, a major 'greenhouse gas', but others such as N_2O, CH_4, CO and CH_3Cl also are generated as direct or indirect products of marine biological activity. Planktonic photosynthesis provides an important sink for CO_2 and its effectiveness is dependent on nutrient controls such as phosphate and nitrate and some trace elements such as iron. Other gases in the marine atmosphere, such as dimethyl sulfide, also have important climatic effects, such as influencing cloud formation.

In the third of the articles, Peter Swarzenski of the US Geological Survey Center for Coastal Geology in St Petersburg, Florida, and his colleagues Reide Corbett from Florida State University, Joseph Smoak of the University of Florida, and Brent McKee of Tulane University, describe the use of uranium–thorium series radionuclides and other transient tracers in oceanography. The former set of radioactive tracers occur naturally in seawater as a product of weathering or mantle emanation and, via the parent–daughter isotope relationships, can provide an apparent time stamp for both water column and sediment processes. In contrast, transient anthropogenic tracers such as the freons or CFCs are released into the atmosphere as a byproduct of industrial/municipal activity. Wet/dry precipitation injects these tracers into the sea where they can be used to track such processes as ocean circulation or sediment accumulation. The use of tracers has been critical to the tremendous advances in our understanding of major oceanic cycles that have occurred in the last 10–20 years. These tracer techniques underpin much of the work in such large-scale oceanographic programmes as WOCE (World Ocean Circulation Experiments) and JGOFS (Joint Global Ocean Flux Study).

The next article is by Raymond Andersen and David Williams of the Departments of Chemistry and of Earth and Ocean Sciences at the University of British Columbia. This is concerned with the opportunities and challenges involved in developing new pharmaceuticals from the sea. Historically, drug discovery programmes have relied on *in vitro* intact-tissue or cell-based assays to screen libraries of synthetic compounds or natural product extracts for pharmaceutically relevant properties. However, modern 'high-throughput screening' methods have vastly increased the numbers of assays that can be performed, such that libraries of up to 100 000 or more chemical entities can now be screened for activity in a reasonable time frame. This has opened the way to exploitation of natural products from the oceans in this context. Many of these marine natural products have no terrestrial counterparts and offer unique opportunities for drug applications. Examples of successful marine-derived drugs are given and the potential for obtaining many more new pharmaceuticals from the sea is clearly demonstrated.

The final article of the book is by Stephen de Mora of the International Atomic Energy Agency's Marine Environment Laboratory in Monaco and is concerned with contamination and pollution in the marine environment. The issues addressed range from industrial and sewage discharges and the effects of elevated nutrients from agricultural runoff in coastal zones to contamination of the deep oceans by crude oil, petroleum products and plastic pollutants, as well as wind-borne materials such as heavy metals. The use of risk assessment and bioremediation methods is reviewed and a number of specific case studies involving such problems as persistent organic pollutants and the use of anti-fouling paints containing organotin compounds are detailed. An overview of the economic and legal considerations relevant to marine pollution is given.

Taken together, this set of articles provides a wide-ranging and authoritative review of the current state of knowledge in the field and a depth of treatment of many of the most important issues relating to chemistry in the marine environment. The volume will be of interest equally to environmental scientists, to chemical oceanographers, and to national and international policymakers concerned with marine pollution and related matters. Certainly it is expected to be essential reading for students in many environmental science and oceanography courses.

Ronald E. Hester
Roy M. Harrison

Contents

Issues in Environmental Science and Technology No. 13
Chemistry in the Marine Environment
© The Royal Society of Chemistry, 2000

Editors

Ronald E. Hester, BSc, DSc(London), PhD(Cornell), FRSC, CChem

Ronald E. Hester is Professor of Chemistry in the University of York. He was for short periods a research fellow in Cambridge and an assistant professor at Cornell before being appointed to a lectureship in chemistry in York in 1965. He has been a full professor in York since 1983. His more than 300 publications are mainly in the area of vibrational spectroscopy, latterly focusing on time-resolved studies of photoreaction intermediates and on biomolecular systems in solution. He is active in environmental chemistry and is a founder member and former chairman of the Environment Group of the Royal Society of Chemistry and editor of 'Industry and the Environment in Perspective' (RSC, 1983) and 'Understanding Our Environment' (RSC, 1986). As a member of the Council of the UK Science and Engineering Research Council and several of its sub-committees, panels and boards, he has been heavily involved in national science policy and administration. He was, from 1991–93, a member of the UK Department of the Environment Advisory Committee on Hazardous Substances and is currently a member of the Publications and Information Board of the Royal Society of Chemistry.

Roy M. Harrison, BSc, PhD, DSc (Birmingham), FRSC, CChem, FRMetS, FRSH

Roy M. Harrison is Queen Elizabeth II Birmingham Centenary Professor of Environmental Health in the University of Birmingham. He was previously Lecturer in Environmental Sciences at the University of Lancaster and Reader and Director of the Institute of Aerosol Science at the University of Essex. His more than 250 publications are mainly in the field of environmental chemistry, although his current work includes studies of human health impacts of atmospheric pollutants as well as research into the chemistry of pollution phenomena. He is a past Chairman of the Environment Group of the Royal Society of Chemistry for whom he has edited 'Pollution: Causes, Effects and Control' (RSC, 1983; Third Edition, 1996) and 'Understanding our Environment: An Introduction to Environmental Chemistry and Pollution' (RSC, Third Edition, 1999). He has a close interest in scientific and policy aspects of air pollution, having been Chairman of the Department of Environment Quality of Urban Air Review Group as well as currently being a member of the DETR Expert Panel on Air Quality Standards and Photochemical Oxidants Review Group, the Department of Health Committee on the Medical Effects of Air Pollutants and Chair of the DETR Atmospheric Particles Expert Group.

Contributors

R.J. Andersen, *Department of Chemistry, 2036 Main Mall, University of British Columbia, Vancouver, British Columbia V6T 1Z1, Canada*
G.R. Bigg, *School of Environmental Sciences, University of East Anglia, Norwich NR4 7TJ, UK*
D.R. Corbett, *Department of Oceanography, Florida State University, Tallahassee, FL 32306, USA*
S.J. de Mora, *Marine Environment Laboratory, International Atomic Energy Agency, 4 Quai Antoine 1er, BP 800, MC 98012, Monaco*
B.A. McKee, *Department of Geology, Tulane University, New Orleans, LA 70118, USA*
W.L. Miller, *Department of Oceanography, Dalhousie University, Halifax, Nova Scotia B3H 4J1, Canada*
J.M. Smoak, *Department of Fisheries and Aquatic Sciences, University of Florida, Gainesville, FL 32653, USA*
P.W. Swarzenski, *US Geological Survey, Center for Coastal Geology, 600 4th Street South, St.Petersburg, FL 33701, USA*
D.E. Williams, *Department of Earth and Ocean Sciences, University of British Columbia, Vancouver, British Columbia V6T 1Z1, Canada*

Introduction and Overview

WILLIAM L. MILLER

1 Introduction

Why does *Chemistry in the Marine Environment* deserve separate treatment within the *Issues in Environmental Science and Technology* series? Is it not true that chemical principles are universal and chemistry in the oceans must therefore simply abide by these well-known laws? What is special about marine chemistry and chemical oceanography?

The long answer to those questions would probably include a discourse on complex system dynamics, carefully balanced biogeochemical cycles, and perhaps throw in a bit about global warming, ozone holes, and marine resources for relevance. The short answer is that marine chemistry *does* follow fundamental chemical laws. The application of these laws to the ocean, however, can severely test the chemist's ability to interpret their validity. The reason for this relates to three things: (1) the ocean is a complex mixture of salts, (2) it contains living organisms and their assorted byproducts, and (3) it covers 75% of the surface of the Earth to an average depth of almost 4000 metres. Consequently, for the overwhelming majority of aquatic chemical reactions taking place on this planet, chemists are left with the challenge of describing the chemical conditions in a high ionic strength solution that contains an unidentified, modified mixture of organic material. Moreover, considering its tremendous size, how can we reasonably extrapolate from a single water sample to the whole of the oceans with any confidence?

The following brief introduction to this issue will attempt to provide a backdrop for examining some marine chemical reactions and distributions in the context of chemical and physical fundamentals. The detailed discussions contained in the chapters that follow this one will provide examples of just how well (or poorly) we can interpret specific chemical oceanographic processes within the basic framework of marine chemistry.

Issues in Environmental Science and Technology No. 13
Chemistry in the Marine Environment
© The Royal Society of Chemistry, 2000

2 The Complex Medium Called Seawater

For all of the millions of years following the cooling of planet Earth, liquid water has flowed from land to the sea. Beginning with the first raindrop that fell on rock, water has been, and continues to be, transformed into planetary bath water as it passes over and through the Earth's crust. Rivers and groundwater, although referred to as 'fresh', contain a milieu of ions that reflect the solubility of the material with which they come into contact during their trip to the sea. On a much grander scale even than the flow of ions and material to the ocean, there is an enormous equilibration continually in progress between the water in the ocean and the rock and sediment that represents its container. Both the low-temperature chemical exchanges that occur in the dark, high-pressure expanses of the abyssal plains and the high-temperature reactions occurring within the dynamic volcanic ridge systems contribute controlling factors to the ultimate composition of seawater.

After all those many years, the blend of dissolved materials we call seawater has largely settled into an inorganic composition that has remained unchanged for thousands of years prior to now. Ultimately, while Na^+ and Cl^- are the most concentrated dissolved components in the ocean, seawater is much more complex than a solution of table salt. In fact, if one works hard enough, every element in the periodic table can be measured as a dissolved component in seawater. In addition to this mix of inorganic ions, there is a continual flux of organic molecules cycling through organisms into the ocean on timescales much shorter than those applicable to salts. Any rigorous chemical calculation must address both.

Salinity and Ionic Strength

The saltiness of the ocean is defined in terms of salinity. In theory, this term is meant to represent the total number of grams of dissolved inorganic ions present in a kilogram of seawater. In practice, salinity is determined by measuring the conductivity of a sample and by calibration through empirical relationships to the International Association of Physical Sciences of the Ocean (IAPSO) Standard Sea Water. With this approach, salinity can be measured with a precision of at least 0.001 parts per thousand. This is fortunate, considering that 75% of all of the water in the ocean falls neatly between a salinity of 34 and 35. Obviously, these high-precision measurements are required to observe the small salinity variations in the ocean.

So, why concern ourselves with such a precise measurement of salinity? One physical consequence of salinity variations is their critical role in driving large-scale circulation in the ocean through density gradients. As for chemical consequences, salinity is directly related to ionic concentration and the consequent electrostatic interactions between dissolved constituents in solution. As salinity increases, so does ionic strength. Because the thermodynamic constants relating to any given reaction in solution are defined in terms of chemical activity (not chemical concentration), high ionic strength solutions such as seawater can result in chemical equilibria that are very different from that defined with thermodynamic constants at infinite dilution. This is especially true of seawater, which contains substantial concentrations of CO_3^{2-}, SO_4^{2-}, Mg^{2+}, and Ca^{2+}. These doubly charged ions create stronger electrostatic interactions than the singly charged ions found in a simple NaCl solution.

Changes in activity coefficients (and hence the relationship between concentration and chemical activity) due to the increased electrostatic interaction between ions in solution can be nicely modeled with well-known theoretical approaches such as the Debye–Hückel equation:

$$\log \gamma_i = -Az_i^2\sqrt{I} \qquad (1)$$

where γ is the activity coefficient of ion i, A is its characteristic constant, z is its charge, and I is the ionic strength of the solution. Unfortunately, this equation is only valid at ionic strength values less than about 0.01 molal. Seawater is typically much higher, around 0.7 molal. Inclusion of additional terms in this basic equation (*i.e.* the extended Debye–Hückel, the Davies equation) can extend the utility of this approach to higher ionic strength and works fine within an ion pairing model for a number of the major and minor ions.[1] Ultimately, however, this approach is limited by a lack of experimental data on the exceedingly large number of possible ion pairs in seawater.

Another approach in the modeling of activity coefficient variations in seawater attempts to take into account all interactions between all species. The Pitzer equations present a general construct to calculate activity coefficients for both charged and uncharged species in solution and form the foundation of the specific interaction model. This complex set of equations, covered thoroughly elsewhere,[2,3] is a formidable tool in the calculation of chemical activity for both charged and uncharged solutes in seawater. Both the ion pairing and the specific interaction models (or a combination of the two) provide valuable information about speciation of both major and trace components in seawater.[4]

Often chemical research in the ocean focuses so intently on specific problems with higher public profiles or greater perceived societal relevance that the fundamental importance of physicochemical models is overlooked. But make no mistake; the inorganic speciation of salts in seawater represents the stage on which all other chemistry in the ocean is played out. These comprehensive inorganic models provide the setting for the specific topics in the following chapters. While these models represent significant advances in the understanding of marine chemistry, seawater, however, is such a complex mixture that on occasion even sophisticated models fail to accurately describe observations in the real ocean. In these cases, the marine chemist is left with empirical descriptions as the best predictive tool. Sometimes this situation arises owing to processes such as photochemistry or biochemical redox reactions that push systems away from equilibrium. Other times it results from the presence of unknown and/or

[1] F.J. Millero and D.R. Schreiber, *Am. J. Sci.*, 1982, **282**, 1508.

[2] K.S. Pitzer, in *Activity Coefficients in Electrolyte Solutions*, ed. K.S. Pitzer, CRC Press, Boca Raton, FL, 1991, p. 75.

[3] F.J. Millero, in *Marine Chemistry: An Environmental Analytical Chemical Approach*, ed. A. Gianguzza, E. Pelizzetti and S. Sammartano, Kluwer, Dordrecht, 1997, p. 11.

[4] F.J. Millero, *Geochim. Cosmochim. Acta*, 1992, **56**, 3123.

uncharacterized compounds. Many of these latter compounds are of biological origin.

Biological Contributions

In sharp contrast to the cool precision of the electrostatic equations used to describe the inorganic interactions discussed above, the study of organic chemistry in the ocean does not enjoy such a clear approach to the evaluation of organic compounds in seawater. There is a boundless variety of both terrestrial and marine organisms that contribute organic compounds to the sea. While their initial contributions may be recognized as familiar biochemicals, much of this material is quickly transformed by microbial and chemical reactions into a suite of complex macromolecules with only a slight resemblance to their precursors. Consequently, the starting point for evaluation of a general approach for organic chemistry in the ocean is a situation where more than half of the dissolved organic carbon (DOC) is contained in molecules and condensates that are not structurally characterized; a mixture usually referred to as humic substances (HS). In other words, for many of the organic reactions in the ocean, we simply do not know the reactants.

Humic substances in the ocean are thought to be long lived and relatively unavailable for biological consumption. They are found at all depths and their average age in the deep sea is estimated in the thousands of years.[5] This suggests that they are resilient enough to survive multiple complete trips through the entire ocean system. The chromophoric (or coloured) dissolved organic matter (CDOM), which absorbs most of the biologically damaging, high-energy ultraviolet radiation (UVR) entering the ocean, is composed largely of HS. Consequently, HS, through its light gathering role in the ocean, protects organisms from lethal genetic damage and provides the primary photon absorption that drives photochemistry in the ocean. Since UVR-driven degradation of CDOM (and HS) both oxidizes DOC directly to volatile gases (primarily CO_2 and CO) and creates new substrate for biological degradation, the degree to which HS is exposed to sunlight may ultimately determine its lifetime in the ocean. Since DOC represents the largest organic carbon pool reactive enough to respond to climate change on timescales relevant to human activity, its sources and sinks represent an important aspect in understanding the relation between ocean chemistry and climate change.

The presence of HS in seawater does more than provide a carbon source for microbes and alter the UV optical properties in the ocean. It can also affect the chemical speciation and distribution of trace elements in seawater. Residual reactive sites within the highly polymerized mixture (*i.e.* carboxylic and phenolic acids, alcohols, and amino groups) can provide binding sites for trace compounds. The chemical speciation of Cu in seawater is a good example of a potentially toxic metal that has a distribution closely linked to that of HS and DOC. A very large percentage of Cu is complexed to organic compounds in seawater and consequently rendered non-toxic to most organisms since the free ion form of Cu

[5] P.M. Williams and E.R.M. Druffel, *Nature*, 1987, **330**, 246.

is usually required for accumulation. One study of Cu in a sewage outfall area within Narragansett Bay, RI, USA shows this effect dramatically.[6] As expected, the highest total Cu concentrations were found in this most impacted area of the estuary. Exactly coincident with high Cu concentrations, the researchers found the lowest Cu toxicity due to high DOC concentrations and increased complexation. Even though specific organic ligands could not be identified, it was clear that the presence of undefined organic compounds had turned a potentially lethal Cu solution into a refuge from toxicity.

The compounds that *are* identifiable in the sea represent a vast array of biochemicals attributable to the life and death of marine plants and animals. They are generally grouped into six classes based on structural similarities: hydrocarbons, carbohydrates, lipids, fatty acids, amino acids, and nucleic acids. Because they represent compounds that can be quantified and understood for their chemical properties and known role in biological systems, a great deal of information has been accumulated over the years about these groups and the specific compounds found within them.[7]

While each *individual* organic compound may exist in exceedingly low concentrations, its presence in solution can be quite important. Organic carbon leaking into solution from the death of organisms can serve as a potential food source for a community of decomposers. Other compounds are intentionally excreted into solution, potentially affecting both biological and chemical surroundings. Certain of these compounds found in marine organisms are unique in their ability to elicit a particular biological or chemical effect. Some biochemicals may serve to attract mates or repel predators and others have the ability to sequester specific required nutrients, in particular, essential trace metals. An excellent example of the ability of small concentrations of biochemicals to significantly impact marine chemistry can be seen in a recent examination of iron speciation in the ocean.[8]

Given the slightly alkaline pH of seawater, and relatively high stability constants for Fe(III) complexes with hydroxide in seawater, it has long been believed that the hydrolysis of Fe(III) represents the main speciation for iron in the ocean. The low solubility of $Fe(OH)_3$ keeps total iron concentrations in the nanomolar range. Consequently, calculations of iron speciation based on known thermodynamic relationships have been extremely difficult to confirm experimentally at natural concentrations. In recent years, the use of ultraclean techniques with electrochemical titrations has turned the idea of a seawater iron speciation dominated by inorganic chemistry on its ear. Working on seawater samples from many locations, several groups[9,10] have shown the presence of a natural organic ligand (also at nanomolar concentrations) that specifically binds to Fe(III). In fact, this ligand possesses conditional stability constants for

[6] W. G. Sunda and A. W. Hanson, *Limnol. Oceanogr.*, 1987, **32**, 537.

[7] J. W. Farrington, 'Marine Organic Geochemistry: Review and Challenges for the Future', *Mar. Chem.*, special issue 1992, **39**.

[8] K. W. Bruland and S. G. Wells, 'The Chemistry of Iron in Seawater and its Interaction with Phytoplankton', *Mar. Chem.*, special issue, 1995, **50**.

[9] E. L. Rue and K. W. Bruland, *Mar. Chem.*, 1995, **50**, 117.

[10] C. M. G. van den Berg, *Mar. Chem.*, 1995, **50**, 139.

association with the ferric ion that are so high ($K_L \approx 10^{20}$ M^{-1}) that it completely dominates the speciation of iron in the ocean. Calculations that include this ligand predict that essentially *all* of the iron in the ocean is organically complexed. In view of the fact that Fe is an essential nutrient and can limit primary productivity in the ocean, the chemistry associated with this Fe ligand represents quite a global impact for such a seemingly insignificant concentration of a very specific organic compound; a compound that was only discovered as a dissolved constituent in seawater within the last 10 years.

3 Spatial Scales and the Potential for Change

As mentioned in the introduction to this chapter, the ocean is enormous. One compilation[11] that includes all of the oceans and adjacent seas puts the volume of seawater on the planet at 1.37×10^9 km^3 covering 3.61×10^8 km^2. The Atlantic, Pacific, and Indian oceans alone contain about 320 million km^3 (or 3×10^{20} litres) of seawater. Consequently, when we consider a ubiquitous chemical reaction in seawater, no matter how insignificant it may seem to our ordinary scale of thinking, its extrapolation to such huge proportions can result in the reaction taking on global significance. Conversely, chemical modifications that create a considerable local impact may be of no consequence when considered in the context of the whole ocean. The sheer size of the ocean forces a unique approach when applying chemical principles to the sea.

Separation of the Elements

Because the ocean spreads continuously almost from pole to pole, there is a large degree of difference in the heating of surface waters owing to varying solar radiation. This causes variations in both temperature (obviously) and salinity (from differential evaporation:precipitation ratios). These variations in heat and salt drive a great thermohaline circulation pattern in the ocean that witnesses cold, salty water sinking in the north Atlantic and in Antarctica's Weddell Sea, flowing darkly through the ocean depths, and surfacing again in the North Pacific; a journey lasting approximately 1000 years. This deep, dense water flows beneath the less dense surface waters and results in a permanent pycnocline (density gradient) at about 1000 metres; a global barrier to efficient mixing between the surface and deep oceans. The notable exceptions to this stable situation are in areas of the ocean with active upwelling driven by surface currents. On a large scale, the ocean is separated into two volumes of water, largely isolated from one another owing to differences in salinity and temperature. As mentioned above, both of these variables will produce changes in fundamental equilibrium and kinetic constants and we can expect different chemistry in the two layers.

Another layering that occurs within the 1000 metre surface ocean is the distinction between seawater receiving solar irradiation (the photic zone) and the dark water below. The sun provides heat, UVR, and photosynthetically active

[11] J. A. Knauss, *An Introduction to Physical Oceanography*, Prentice Hall, Englewood Cliffs, NJ, 1978, p. 2.

radiation (PAR) to the upper reaches of the ocean. Heat will produce seasonal pycnoclines that are much shallower than the permanent 1000 metre boundary. Winter storms limit the timescale for seasonal pycnoclines by remixing the top 1000 metres on roughly a yearly basis. Ultraviolet radiation does not penetrate deeply into the ocean and limits photochemical reactions to the near surface (metres to tens of metres depending on the concentration of CDOM). The visible wavelengths that drive photosynthesis penetrate deeper than UVR but are still generally restricted to the upper hundred metres.

At almost any location in the open ocean, the underlying physical structure provides at least three distinct volumes of water between the air–sea interface and the bottom. This establishes the potential for vertical separation of elements into distinct chemical domains that occupy different temporal and spatial scales. In fact, the biological production of particles in the photic zone through photosynthesis acts to sequester a wide variety of chemical elements through both direct incorporation into living tissue and skeletal parts and the adsorption of surface reactive elements onto particles. Nutrients essential to marine plant growth like N, P, Si, Fe, and Mn are stripped from the photic zone and delivered to depth with particles. While most of the chemicals associated with particles are recycled by microbial degradation in the upper 1000 metres, some percentage drop below the permanent pycnocline and return to the dissolved components of the deep ocean through microbial degradation and chemical dissolution. This flux of particles from the surface ocean to deeper waters leads to vertical separation of many chemical elements in the ocean.

The redistribution of essential biological elements away from where they are needed for photosynthesis sets up an interesting situation. Marine plants, limited to the upper reaches of the ocean by their need for light, are floating in a seawater solution stripped of many of the chemicals required for growth. Meanwhile, beneath them, in the deep ocean layers, exists the largest storehouse of plant fertilizer on the planet; a reservoir that grows ever larger as it ages. The mechanisms and rates of this particle-driven, chemical separation of the 'fuel and the fire' are more closely examined by P. W. Swarzenski and co-authors later in this book.

Diversity of Environments

Along with the great depth that leads to the vertical separation of water masses with different density, the horizontal distribution of surface seawater across all climates on Earth leads to a diversity of environments that is unlike any terrestrial system. While terrestrial ecosystems often offer up physical barriers to migration, the oceans are fluid and continuous. The mountains and trenches found on the ocean floor present little or no barrier to organisms that have evolved for movement and dispersal of offspring in three-dimensional space. With enough time and biological durability, organisms thriving in any part of the ocean could potentially end up being transported to any other part of the ocean. The demarcations between different marine environments are often gradual and difficult to define.

Ecological distinctions are easy to recognize when considering the ocean floor:

muddy, sandy, or rocky bottoms result in very different benthic ecosystems. In the majority of the ocean, however, organisms face pelagic distinctions that are defined by varying physical and chemical characteristics of the solution itself. Temperature is an obvious environmental factor. Most arctic organisms do not thrive in tropical waters, although they may have closely related species that do. A more subtle result of temperature variation involves the solubility of calcium carbonate. The fact that calcium carbonate is less soluble in warm water than in cold dictates the amount of energy required by plants and animals to build and maintain calcium carbonate structures. This simple chemistry goes a long way toward explaining the tropical distribution of massive coral reefs. Salinity, while showing little variation in the open ocean, can define discrete environments where rivers meet the sea. Chemical variations much more subtle than salinity can also result in finely tuned ecological niches, some as transient as the sporadic events that create them.

In the deep sea, entire ecosystems result from the presence of reduced compounds like sulfur and iron in the water. These chemicals, resulting from contact between seawater and molten rock deep within the Earth, spew from vents within the superheated seawater. Their presence fuels a microbial population that serves as the primary producers for the surrounding animal assemblage, the only known ecosystem not supported by photosynthesis. Both the reduced elements and the vents themselves are transient. Sulfide and Fe(II) are oxidized and lost as the hot, reducing waters mix with the larger body of oxygenated water. Vents are periodically shut down and relocated tens to hundreds of kilometres away by volcanic activity and shifting of crustal rock. Yet, these deep sea organisms have the intricate biochemistry to locate and exploit chemical anomalies in the deep ocean.

Variable chemical distributions of specific elements in the ocean promote finely tuned biological systems capable of exploiting each situation presented. For example, the addition of Fe to open ocean ecosystems that are starved of this micronutrient will cause population shifts from phytoplankton species that thrive in low iron environments to those with higher Fe requirements. This shift in plant speciation and growth can alter the survival of grazer populations and their predators further up the food chain. It is important to note from this example that chemical changes in the nanomolar range are certainly capable of altering entire marine ecosystems.

In short, seemingly small chemical and physical gradients within seawater can dictate the success or failure of organisms that possess only subtle differences in biochemical machinery and will push marine ecosystems towards increased biodiversity. The presence of a specific set of organisms in seawater will produce a distinct chemical milieu via incorporation of required elements and excretion of others. Salmon, returning from the ocean to spawn, can identify the set of chemicals specific to the streams and rivers of their birth. The biochemistry of marine organisms is very often finely evolved to exploit almost imperceptible changes in ocean chemistry. Many other biochemical adaptations have resulted in response to the intense competition among organisms to exploit these tiny changes in their environment. Almost certainly, there are innumerable examples that man has not yet even identified. Many of these specific compounds are being discovered and their sources and prospects for exploitation are examined in the chapter in this book by R.J. Andersen and D.E. Williams.

Impacts

Because their survival often depends directly on the ability to detect and respond to infinitesimal changes in seawater chemistry, many marine organisms are extremely sensitive to the presence of man-made contaminants in the ocean. As mentioned above, it only requires nanomolar concentrations of Fe to change entire marine ecosystems and potentially alter the chemical distribution of all elements integral to the resulting biological processes. These intricate changes may not be easily observable. The truth is, contamination may have already altered the ocean in subtle ways that we currently know nothing about. The more obvious examples of man's impact on the ocean can be seen on smaller scales in areas closer to anthropogenic activity, namely the coastal zone.

Our most vivid examples of man's impact on marine systems often result from catastrophic episodes such as oil spills and the visible results from marine dumping of garbage. Oil drenched seabirds, seashores littered with dead fish, and medical refuse on public beaches are the images that spring to mind when considering marine pollution. While these things do represent the worst local impact that man has been able to impose on the ocean, they probably do not represent the largest threat to marine systems. Non-pointsource pollution such as terrestrial runoff of fertilizers and pesticides, discharge of long-lived industrial chemical pollutants, daily spillage of petroleum products from shipping activities, and increasing concentrations of atmospheric contaminants all reflect man's chronic contribution to ocean chemistry. These activities have the potential to accumulate damage and affect the natural chemical and biological stasis of the ocean. A subsequent chapter in this book by S. J. de Mora provides many more details on the chronic and episodic modifications of marine chemistry that can result from man's activities.

As pointed out earlier, it is difficult to effect chemical change over the entire ocean owing to its great size. Consequently, changes to the whole ocean system are usually slow, only observable over hundreds to thousands of years. This is not to say that long-term chemical changes cannot result from man's activities. Atmospheric delivery of anthropogenic elements can spread pollutants to great distances and result in delivery of material to large expanses of the ocean. Outside of the obvious impact of natural phenomena like large-scale geological events and changes in solar insolation, the exchange of material between the ocean and atmosphere represents one of the few mechanisms capable of producing oceanic changes on a global scale. Examination of the exchange of material between marine and atmospheric chemistry forces the collaboration of two disciplines: oceanography and atmospheric science. Recent scientific enterprise directed at the understanding of climate change and man's potential role in that change has led to a closer collaboration between these two disciplines than ever before. A subsequent chapter in this book by G. R. Bigg goes into detail as to the workings of ocean–atmosphere exchange.

Part of the requirement for interdisciplinary efforts in ocean–atmosphere exchange can be seen in a *qualitative* way by examining the dimethyl sulfide (DMS) story.[12] It should be noted that development of many *quantitative* aspects of this story are still on the drawing board and once these details are resolved, the future telling of this story could very easily have a different plot and finale. Regardless of the eventual details, the original DMS story reveals a glimpse into the complex processes, reciprocal impacts, and feedback loops that must be unearthed to understand the exact role of ocean–atmosphere exchange in climate change.

The DMS story begins with the observation that in remote areas of the open ocean this trace gas is found both in the atmosphere and in surface waters with the relative concentrations indicating an oceanic source. The intriguing part of the story emerges when one considers the source of DMS in the ocean and its eventual role in the remote atmosphere. Phytoplankton are responsible for the precursors for DMS production in the surface ocean, where it fluxes into the troposphere. Through redox chemistry in the atmosphere, it appears that DMS is capable, at least in part, of supplying the sulfate aerosols that serve as cloud condensation nuclei. In other words, an organism that directly depends on solar irradiance for its survival is the sole supplier of a compound that makes clouds. This formation of clouds, in turn, changes the intensity and spectral quality of light reaching the surface ocean. It is well known that phytoplankton growth, with nutrients available, is directly regulated by the quantity and quality of sunlight. Do phytoplankton population dynamics have a feedback mechanism with cloud formation through the formation of DMS?

In another twist to the story, we know that many biological systems, with all other growth parameters being equal, will operate at increased rates when warmed. It is also known that white clouds have a higher albedo than ocean water, thereby reflecting more sunlight back toward space. Does it then follow that global warming will increase phytoplankton growth rates and result in enhanced global DMS formation? Will this new elevated DMS flux result in more clouds over the ocean? If so, will the increased albedo cool the atmosphere and serve as a negative feedback to global warming?

With the purposeful omission of the details in the DMS story as told here, it is not possible to answer these questions. It is, however, possible to imagine that the distribution and chemistry of a simple biogenic sulfur gas can have global implications. Additionally, there are biogenic and photochemical sources of other atmospherically significant trace gases in the ocean. Carbon monoxide, carbonyl sulfide, methyl bromide, methyl iodide, and bromoform[13,14] all have oceanic sources to the atmosphere. In the end, it appears that this feedback between processes in marine surface waters and atmospheric chemistry is an integral part of climate control. Through this connection, it is quite possible that man's impact on the oceans can spread far beyond local events.

[12] R.J. Charlson, J.E. Lovelock, M.O. Andreae and S.G. Warren, *Nature*, 1987, **326**, 655.

[13] R.M. Moore and R. Tokarczyk, *Global Biogeochem. Cycles*, 1993, **7**, 195.

[14] P.S. Liss, A.J. Watson, M.I. Liddicoat, G. Malin, P.D. Nightingale, S.M. Turner and R.C. Upstill-Goddard, in *Understanding the North Sea System*, ed. H. Charnock, K.R. Dyer, J. Huthnance, P.S. Liss, J.H. Simpson and P.B. Tett, Chapman and Hall, London, 1993, p. 153.

4 Summary

The field of chemical oceanography/marine chemistry considers many processes and concepts that are not normally included in a traditional chemical curriculum. While this fact makes the application of chemistry to the study of the oceans difficult, it does not mean that fundamental chemical principles cannot be applied. The chapters included in this book provide examples of important chemical oceanographic processes, all taking place within the basic framework of fundamental chemistry. There are three principal concepts that establish many of the chemical distributions and processes and make the ocean a unique place to practice the art of chemistry: (1) the high ionic strength of seawater, (2) the presence of a complex mixture of organic compounds, and (3) the sheer size of the oceans.

The physicochemical description of seawater must include the electrostatic interactions between a multitude of different ions dissolved in the ocean. This high ionic strength solution provides the matrix that contains and controls all other chemical reactions in the sea. Much of the dissolved organic carbon that is added to this milieu by biological activity is composed of a mixture of molecules and condensates that are not yet identified, making a description of their chemistry difficult. The identifiable organic compounds, while almost always present at very low concentrations, can greatly affect the distribution of other trace compounds and even participate in climate control via feedback to primary production and gas exchange with the atmosphere.

A combination of water mass movement and the biological formation of particles that strip chemicals from solution causes the physical separation of many elements into vertical zones. Given the great depth and expanse of the ocean, a spatial and temporal distribution of chemicals is established that controls many biological and chemical processes in the sea. These spatial gradients of chemical and physical seawater parameters encourage a diversity of organisms that are sensitive to remarkably small changes in their chemical surroundings. While the impact on the ocean by man's activities is often local in effect, the combination of a carefully poised chemistry, a population of chemically sensitive organisms, and the continued contribution of anthropogenic products through atmospheric transport sets up the possibility of impact on a global scale. The chapters contained in this book are just a few examples of the important areas of marine chemistry that require understanding and evaluation in order to fully grasp the role of the oceans within our planetary system.

The Oceans and Climate

GRANT R. BIGG

1 Introduction

The ocean is an integral part of the climate system. It contains almost 96% of the water in the Earth's biosphere and is the dominant source of water vapour for the atmosphere. It covers 71% of the planet's surface and has a heat capacity more than four times that of the atmosphere. With more than 97% of solar radiation being absorbed that falls on the surface from incident angles less than 50° from the vertical, it is the main store of energy within the climate system.

Our concern here is mainly with the chemical interaction between the ocean and atmosphere through the exchange of gases and particulates. Through carbonate chemistry the deep ocean is a major reservoir in the global carbon cycle, and so can act as a long-term buffer to atmospheric CO_2. The surface ocean can act as either a source or sink for atmospheric carbon, with biological processes tending to amplify the latter. Biological productivity, mostly of planktonic life-forms, plays a major role in a number of other chemical interactions between ocean and atmosphere. Various gases that are direct or indirect products of marine biological activity act as greenhouse gases once released into the atmosphere. These include N_2O, CH_4, CO and CH_3Cl. This last one is also a natural source of chlorine, the element of most concern in the destruction of the ozone layer in the stratosphere.

Other, sulfur-related, products of marine biological processes ultimately contribute to production of cloud condensation nuclei (CCN). The physical loss of salt particles to the atmosphere, particularly during wave-breaking, adds to the atmospheric supply of CCN. The oceanic scavenging of atmospheric loadings of some particulate material is also important in this chemical exchange between ocean and atmosphere. Thus nitrates and iron contained in atmospheric dust are fertilizers of marine productivity, and so can potentially act as limiting factors of the biological pump's climatic influence.

Thus the atmospheric component of the planet's radiation budget is strongly modulated by the indirect effects of oceanic gas and particle exchange. As will be

Issues in Environmental Science and Technology No. 13
Chemistry in the Marine Environment
© The Royal Society of Chemistry, 2000

seen in the discussion of feedback processes, altering the radiation budget can have profound impacts on all other aspects of the climate system.

There are also much longer timescales of chemical interaction between the ocean and climate system. These are beyond the scope of this chapter but worth identifying for completeness. The chemical weathering of land surfaces is a mechanism by which changes in the atmospheric concentration of CO_2 can occur over millions of years. For example, slow erosion of the mountain ranges uplifted over the past 20 million years, such as Tibet, the Rocky Mountains and the Alps, sequesters atmospheric CO_2 in the ocean through the run-off of the dissolved carbonate products of weathering.[1] Water and other climatically active compounds are also recycled from the ocean into the atmosphere through tectonic processes. As oceanic plates are subducted under continental crust at destructive plate margins, such as along the west coast of South America, trapped seawater, and its salts, will boil off to become part of the molten crustal matrix that is re-injected into the atmosphere by volcanic activity. These atmospheric inputs can be climatically active, and the whole process helps to maintain the composition of oceanic salinity over geological timescales.[2]

Physical Interaction

While this chapter is mainly concerned with the chemical interactions between ocean and atmosphere, a few words need to be said about the physical interactions, because of their general importance for climate. The main physical interaction between the ocean and atmosphere occurs through the exchange of heat, water and momentum,[3] although the presence of sea-ice acts to reduce all of these exchanges to a greater or lesser extent.

Momentum is mostly transferred from the atmosphere to the ocean, having the effect of driving the ocean circulation through the production of a wind-driven flow. Of course, the resultant flow carries heat and water, so contributing to fluxes of these quantities to the atmosphere in ways that would not have occurred without the establishment of the wind-driven circulation in the first place.

Heat is transferred in both directions, affecting the density of each medium, and thus setting up pressure gradients that drive circulation. The ocean radiates infrared radiation to the overlying atmosphere. This is a broadly similar flux globally as it depends on the fourth power of the absolute temperature. In contrast, the amount returned to the ocean through absorption and re-radiation by, particularly, tropospheric water vapour is more variable. Evaporation from the ocean surface, directly proportional to the wind speed as well as the above-water humidity gradient, transfers large, and variable, amounts of latent heat to the atmosphere. This does not warm the atmosphere until condensation occurs, so may provide a means of heating far removed from the source of the original vapour. Zones of concentrated atmospheric heating are also possible by this mechanism, leading to tropical and extra-tropical storm formation. Conduction

[1] M. E. Raymo, *Paleoceanography*, 1994, **9**, 399.

[2] K. B. Krauskopf and D. K. Bird, *Introduction to Geochemistry*, McGraw-Hill, New York, 3rd edn., 1995, ch. 21, p. 559.

[3] G. R. Bigg, *The Oceans and Climate*, Cambridge University Press, Cambridge, 1996, ch. 2, p. 33.

and turbulent exchange also directly transfer heat from the warmer medium, again in proportion to the wind speed. This tends to be much smaller in magnitude than either of the other mechanisms. Latent heat transfer is thus the most temporally and geographically variable heat exchange process, heating the atmosphere at the ocean's expense. Anomalous heating or cooling of the atmosphere over regions of the ocean can lead to atmospheric circulation changes, which in turn can feed back to the maintenance (or destruction) of the originating oceanic anomaly. The El Niño[4] phenomenon in the Pacific is linked to such interactions, as is the North Atlantic Oscillation.[5]

As part of the process of latent heat transfer, water vapour is added to the atmosphere. This not only leads to atmospheric heating through the release of latent heat, but also to cloud formation and maintenance of the natural greenhouse effect through the replenishment of atmospheric water vapour. In exchange, water is added to the surface of the ocean via precipitation, run-off from rivers and melting of icebergs. The local combination of evaporation and addition of fresh water can alter the ocean's surface density considerably. The ocean circulation is a combination of (i) the wind-induced flow and (ii) a larger-scale, deeper-reaching thermohaline circulation, the latter set up by changes in temperature and salinity, and hence density, on both global and regional scales. Altering the surface density regionally could thus have large repercussions for the global ocean circulation, and hence the manner in which the ocean contributes to the climate. Decreasing the salinity of the northern North Atlantic, for example, could significantly slow the meridional overturning circulation, or Conveyor Belt,[6] within the whole Atlantic, which, in turn, means slowing, cooling and alteration of the path of the Gulf Stream extension across the North Atlantic. This would have major climatic effects.[7] We will return to such processes later in this chapter.

The Mechanics of Gas Exchange

The fundamental control on the chemical contribution of the ocean to climate is the rate of gas exchange across the air–sea interface. The flux, F, of a gas across this interface, into the ocean, is often written as

$$F = k_T(C_a - C_s) \tag{1}$$

where C_a and C_s are the respective concentrations of the gas in question in the atmosphere and as dissolved in the ocean, and k_T is the transfer velocity. Sometimes this difference is expressed in terms of partial pressures[8]—in the case of the water value this is the partial pressure that would result if all the dissolved gas were truly in the gaseous state, in air at 1 atmosphere pressure. For gases that

[4] S. G. H. Philander, *El Niño, La Niña and the Southern Oscillation*, Academic Press, New York, 1990, ch. 1, p. 9.

[5] J. M. Wallace and D. S. Gutzler, *Mon. Weather Rev.*, 1981, **109**, 784.

[6] G. R. Bigg *The Oceans and Climate*, Cambridge University Press, Cambridge, 1996, ch. 1, p. 1.

[7] S. Manabe and R. J. Stouffer, *Nature*, 1995, **378**, 165.

[8] F. Thomas, C. Perigaud, L. Merlivat and J.-F. Minster, *Philos. Trans. R. Soc. London, Ser. A*, 1988, **325**, 71.

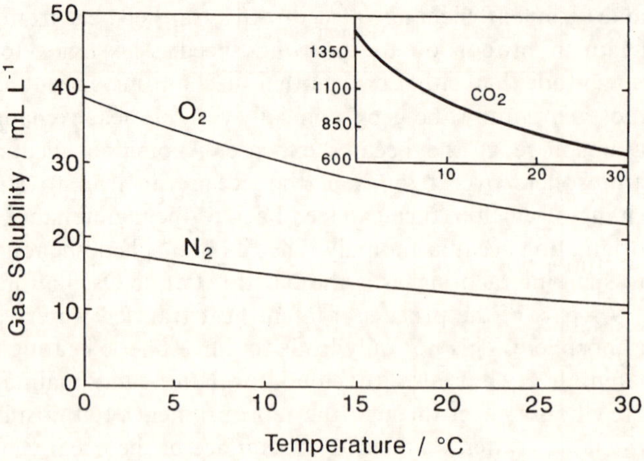

Figure 1 The solubility of the principal atmospheric gases in seawater, as a function of temperature. Units are millilitres of gas contained in a litre of seawater of salinity 35 psu, assuming an overlying atmosphere purely of each gas. Note that salinity is defined in terms of a conductivity ratio of seawater to a standard KCl solution and so is dimensionless. The term 'practical salinity unit', or psu, is often used to define salinity values, however. It is numerically practically identical to the old style unit of parts per thousand by weight

are created through marine biological activity, C_s is generally much larger than C_a so that the net flux towards the atmosphere is directly dependent on the oceanic production rate of the gas. However, if a gas has a large atmospheric concentration, or the ocean can act as a sink for the gas, as with CO_2, then we need to consider the solubility of our gas more carefully, as it is this that will determine $(C_a - C_s)$. For gases that are chemically inert in seawater the solubility is essentially a weak function of molecular weight. Oxygen is a good example of such a gas, although its oceanic partial pressure can be strongly affected by biological processes. For gases like CO_2, however, which have vigorous chemical reactions with water (as we will see in the next section), the solubility is much increased, and has a different temperature dependence. For chemically inert gases the solubility decreases by roughly a third in raising the water's temperature from 0 °C to 24 °C, but for a reactive gas this factor depends on the relative reaction rates. Thus, for CO_2 the solubility more than halves over this temperature range, from 1437 mL L^{-1} to 666 mL L^{-1} (Figure 1).

The other major factor controlling gas exchange is the transfer velocity, k_T. This represents the physical control on exchange through the state of the interface and near-surface atmosphere and ocean.[9] A calm sea, and stable air, will only allow slow exchange because the surface air mass is renewed infrequently and there is largely only molecular diffusion across the interface in these conditions. In very calm conditions the presence of surfactants slows this diffusion even further.[10] Bigger molecules thus have lower values of k_T in low-wave sea states, because diffusion occurs more slowly. By contrast, rough seas and strong winds allow frequent renewal of the surface air, and bubble formation during wave-breaking actively bypasses the much slower molecular diffusion of gas.[11] The molecular size of the gas will also be less important in this strongly physically controlled regime. An abrupt change in transfer rate can be expected when the sea state crosses the transition to breaking waves (Figure 2). Both bulk chemical

[9] P.S. Liss, A.J. Watson, M.I. Liddicoat, G. Malin, P.D. Nightingale, S.M. Turner and R.C. Upstil-Goddard, *Philos. Trans. R. Soc. London, Ser. A*, 1993, **343**, 531.

[10] R. Wanninkhof and W.R. McGillis, *Geophys. Res. Lett.*, 1999, **26**, 1889.

[11] D.M. Farmer, C.L. McNeil and B.D. Johnson, *Nature*, 1993, **361**, 620.

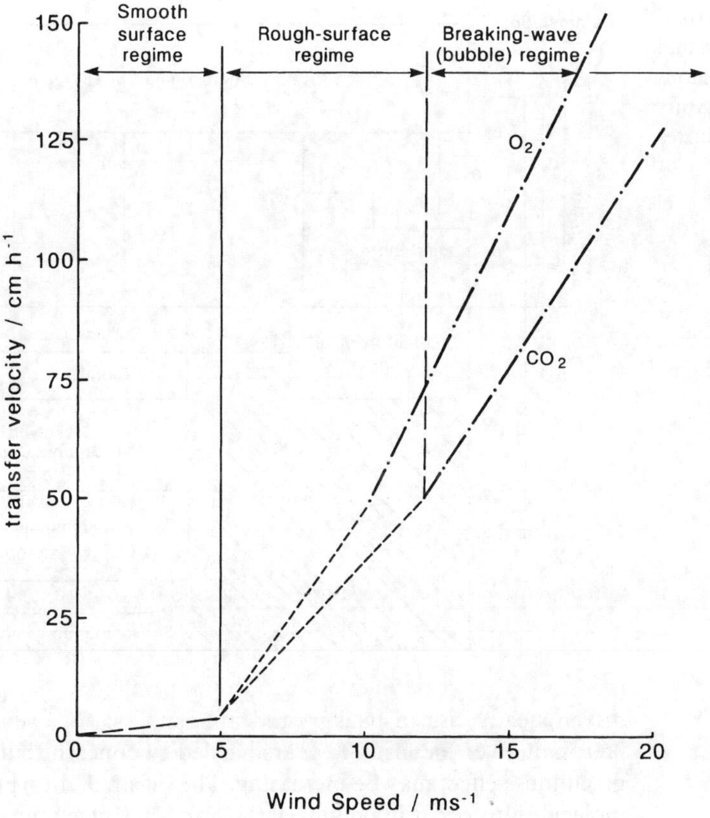

Figure 2 Variation of the gas transfer velocity with wind speed. The units of transfer velocity are equivalent to the number of cm of the overlying air column entering the water per hour (Taken from Bigg,[28] with permission of Cambridge University Press)

measures of this exchange and micrometeorological-based eddy correlation techniques[12] show similar rates of change of k_T with wind speed, but they differ in detail, with the eddy correlation technique tending to give somewhat higher rates of exchange.

A further factor affecting k_T is the air–sea temperature difference. When the sea is colder than the air above it, the enhanced solubility of the gas in the water (relative to the air temperature) tends to increase k_T. This will occur in summer in sub-polar waters and over upwelling regions. The opposite is also found, and much of the ocean equatorward of 45° latitude is colder than the overlying air for much of the year. However, air–sea temperature differences are generally less than 2–3 °C so that this effect results in a less than 10% modulation of k_T on average.

2 Oceanic Gases and the Carbon Cycle

Carbon dioxide is a major greenhouse gas within the atmosphere. Water vapour is a greater contributor to the natural greenhouse effect (55–70% of the total radiative absorption compared to CO_2's 25%). However, the large inherent variability in atmospheric water vapour compared to the anthropogenically

[12] H. Dupuis, P. K. Taylor, A. Weill and K. Katsaros, *J. Geophys. Res.*, 1997, **102**, 21115.

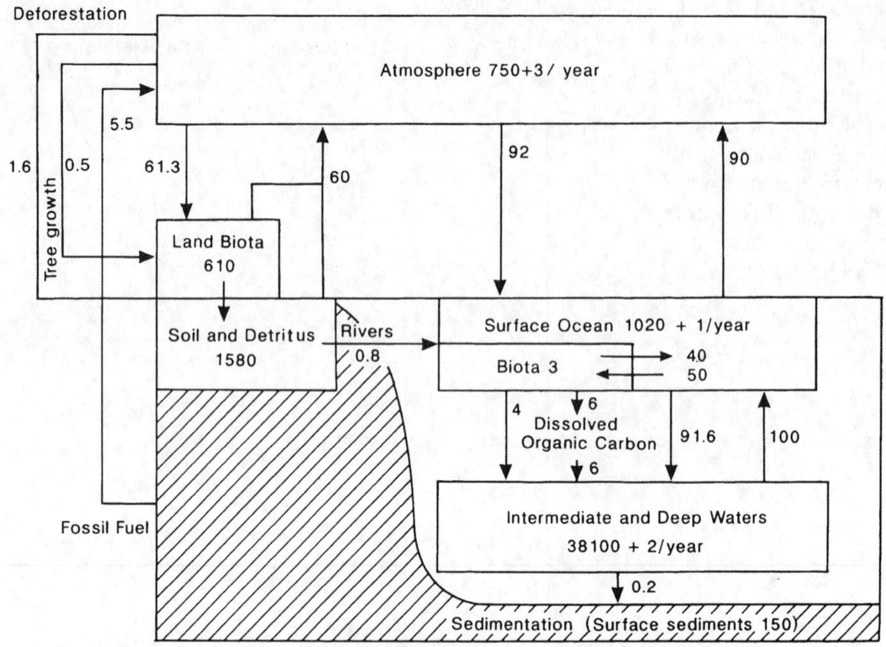

Figure 3 Global carbon reservoirs and annual fluxes.[13] Units are gigatons of carbon in the reservoirs and Gt C yr^{-1} for fluxes

driven steady rise in background atmospheric CO_2 levels from 280 ppmv to 360 ppmv over the last 200 years has led to concern that the magnitude of the greenhouse effect may be increasing. The infrared absorption bands of the CO_2 molecule also occur in regions of the Earth's electromagnetic spectrum where at present moderate amounts of available energy escape to space.

The largest reservoir of available carbon in the global carbon cycle, however, is in the deep ocean, below the thermocline (Figure 3). This is the part of the ocean that has essentially no thermal or dynamical link to direct atmospheric forcing. The depth of the temperature barrier of the thermocline varies geographically and temporally but the deep ocean can roughly be taken to be the entire ocean deeper than 500 m from the surface. Here is stored the end results of the oceanic carbonate chemistry, discussed below. As the overturning, or renewal, timescale of the ocean is of the order of 1000 years, this deep reservoir is essentially isolated from short-term changes to the remainder of the cycle. Smaller reservoirs, but still larger than that in the atmosphere, are found in the upper ocean and the terrestrial biosphere. The upper ocean reservoir has both a chemical and a biological component. While small elements of each of these surficial reservoirs are sequestered into other reservoirs, 5–10% is recycled into the atmosphere each year. Thus both the upper ocean and the terrestrial biosphere have the capacity to interact, subject to a relatively small time lag, with anthropogenically driven atmospheric change to CO_2. As the focus here is on the oceanic involvement with the carbon cycle, mechanisms to significantly alter the biological pump are

[13] D. Schimel, D. Alves, I. Enting and M. Heimann, in *Climate Change 1995*, ed. J. T. Houghton, L. G. Meira Filho, B. A. Callander, N. Harris, A. Kattenberg and K. Maskell, Cambridge University Press, Cambridge, 1996, ch. 2, p. 65.

considered below. This mostly involves ways to alter primary productivity by removing existing trace element controls such as nitrate or iron limitation. These controls are very different in different oceanic regimes: coastal waters have limiting light levels, but excesses of nitrates and iron due to direct input from river run-off or atmospheric deposition; in contrast, open ocean waters may have limits in one nutrient or another depending on the regional physical oceanography.

The ocean's contribution to the carbon cycle has evolved over time, and still changes with the growth and decline of glaciation. However, the deep component of the cycle can also have climatic consequences. If the exchange of carbon shown in Figure 3 is severed through changes to the physical overturning of the ocean as a whole, or a substantial basin, this disconnection of the deep and upper ocean reservoirs can lead to significant climatic change.

Carbonate Chemistry

The basic reason for the ocean being a major sink for CO_2 lies in the reaction of the gas with water, and subsequent anion breakdown:

$$CO_2(gas) + H_2O \rightleftharpoons H^+ + HCO_3^- \rightleftharpoons 2H^+ + CO_3^{2-} \qquad (2)$$

The component reactions in eqn. (2) are very fast, and the system exists in equilibrium. Additional carbon dioxide entering the sea is thus quickly converted into anions, distributing carbon atoms between the dissolved gas phase, carbonate and bicarbonate ions. This storage capacity is clear when the apparent equilibrium constants for the two reactions in eqn. (2) are examined, namely

$$K_1 = \frac{a_{H^+}[HCO_3^-]}{[CO_2]} \qquad (3)$$

for the gas to bicarbonate equilibrium (where $[CO_2]$ is the concentration of the dissolved gas and a_{H^+} is the activity of the hydrogen ion), and

$$K_2 = \frac{a_{H^+}[CO_3^{2-}]}{[HCO_3^-]} \qquad (4)$$

for the bicarbonate to carbonate equilibration. Note that these are different from standard thermodynamic equilibrium constants because of the difficulty in measuring some of the more usual quantities in seawater.[14] These constants depend on temperature, pressure and salinity, most importantly increasing for increasing temperature and pressure.[15] At Standard Temperature and Pressure, and a salinity of 35 psu, K_1 is several orders of magnitude greater than K_2 ($K_1 = 1 \times 10^{-6}$ and $K_2 = 7.69 \times 10^{-10}$) so most carbon is in the intermediate, bicarbonate, reservoir of reaction sequence (2).

More CO_2 can actually be absorbed chemically into the ocean than the above reaction sequence suggests. Terrestrial weathering of rocks containing carbonate, such as limestone, and subsequent aerial or riverine transport, means that the ocean is enriched in carbonate. Keeping K_1 and K_2 constant implies, through eqns. (3) and (4), that enhancing the oceanic $[CO_3^{2-}]$ leads to a greater level of

[14] C. Merbach, C. H. Culberson, J. E. Hawley and R. M. Pytkowicz, *Limnol. Oceanogr.*, 1973, **18**, 897.
[15] W. S. Broecker and T.-H. Peng, *Tracers in the Sea*, Eldigio Press, New York, 1982, ch. 3, p. 110.

oceanic dissolved CO_2. It is also worth noting that seawater's pH is affected by CO_2 dissolution, because hydrogen ions are released in both parts of eqn. (2). Thus, if more carbon is pumped into the system, the greater is the ratio of bicarbonate to carbonate ions, and the more hydrogen ions there are in solution. Thus the pH falls. However, as the bicarbonate to carbonate reaction is so fast (K_2) the carbonate system acts as a pH buffer for the oceans.[16]

Another factor in this reaction sequence is also subject to external modification, namely, moderation of the basic oceanic dissolution of CO_2 through temperature dependence of its solubility, S. The latter is defined as:

$$S = \frac{[CO_2]}{P_a(CO_2)} \tag{5}$$

where $P_a(CO_2)$ is the partial pressure of CO_2 gas in air equilibrated with a particular sample of seawater. Combining eqns. (3), (4) and (5) gives:

$$P_a(CO_2) = \frac{a_{H^+}{}^2[CO_3{}^{2-}]}{K_1K_2S} \tag{6}$$

Relative changes in the components of eqn. (6) mean that $P_a(CO_2)$ should decrease by about a factor of three between 25 °C and 0 °C.[15] However, atmospheric concentrations of CO_2 are essentially uniform, so tropospheric mixing clearly acts fast enough for this potential poleward gradient to be absent.

The Biological Pump

Oceanic biology is a sink for atmospheric CO_2 because of the involvement of the aqueous form of this gas in planktonic photosynthesis. This complex process can be summarized by

$$nCO_2 + nH_2O \xrightarrow{\text{visible light}} nO_2 + (CH_2O)_n \tag{7}$$

where $(CH_2O)_n$ represents a general carbohydrate. The reverse of this process, the absorption of O_2 leading to the release of CO_2, is known as respiration. Different species preferentially absorb different wavelengths of visible light during photosynthesis, and have maximal growth at different temperatures.[17] They can also tolerate more or less intensity of radiation. Thus maximum photosynthesis occurs below the surface and to a varying degree geographically, depending on the environmental conditions and the species distribution.

Other limitations on phytoplankton growth are chemical in nature. Nitrogen, in the form of nitrate, nitrite and ammonium ions, forms a basic building material of a plankton's cells. In some species silicon, as silicate, takes on this role. Phosphorus, in the form of phosphate, is in both cell walls and DNA. Iron, in the form of Fe(III) hydroxyl species, is an important trace element. Extensive areas of the mixed layer of the upper ocean have low nitrate and phosphate levels during

[16] J. Wright and A. Colling, *Seawater: its Composition, Properties and Behaviour*, Pergamon Press, Oxford, 1995, ch. 6, p. 85.

[17] T. R. Parsons, M. Takahashi and B. J. Hargrave, *Biological Oceanographic Processes*, Pergamon Press, Oxford, 1984, ch. 3, p. 65.

periods of planktonic growth. Thus the lack of availability of these species, collectively known as nutrients, can limit phytoplankton growth, and hence marine CO_2 absorption. However, large areas of the ocean have high nutrient levels, but low productivity. These include the Southern Ocean and part of the equatorial seas. Here it is thought that iron is the limiting nutrient,[18] combined, in cold-water environments, with light limitation. Iron is in short supply in the ocean because of the poor dissolution of particles or colloidal material, which is the most common state of marine iron. Field experiments and the marine after-effects of the eruption of Mt. Pinatubo in 1991 have supported this so-called 'iron' hypothesis.[19]

Atmospheric inputs of the various potentially limiting nutrients are considerable and may significantly affect the primary production of the ocean.[20] While globally, and in the long-term, nitrogeneous input from the atmosphere is unlikely to be a major source of planktonic nitrogen, local and short-term effects could be signficant. The major source of oceanic iron, however, is atmospheric. The atmosphere delivers some 30 times more iron to the ocean than rivers, and the sea floor also seems to be a negligible contributor to the overlying water column. Areas far from land, or isolated from the airborne trajectories of major sources of atmospheric dust, such as the Southern Ocean, may therefore be fundamentally limited in their productivity by the lack of iron.[21]

Geographical Variation

The geographical distribution of the mean annual net marine flux of CO_2 to the atmosphere[22] shows the combined influence of physical, chemical and biological effects on marine uptake of carbon (Figure 4). Polar waters tend to have net increases in levels of CO_2 as a result of the united effects of both enhanced solubility and biological production. The sub-tropical oceans are close to a state of equilibrium with the atmosphere, because phytoplankton production is limited by the weak winter upwelling of nutrients. Some such regions, nearby eastern coasts, however, show high levels of CO_2 input to the atmosphere, consistent with oceanic production of CO_2 rather than absorption. This is also visible in equatorial regions. In both cases, such high values occur because water upwells from deeper in the ocean, carrying water that has been at a higher pressure. More carbon can exist as carbonate and bicarbonate ions at the greater pressures at depth, but as the upwelled water's pressure reduces, reaction (2) is pushed to the left and CO_2 is formed. This effect outweighs the carbon draw-down associated with the considerable phytoplankton production in these areas caused by the continual upwelling of nutrients. The net effect is for the ocean to be a source of CO_2 for the atmosphere in these areas.

[18] J. H. Martin and S. E. Fitzwater, *Nature*, 1988, **331**, 341.

[19] A. J. Watson and P. S. Liss, *Philos. Trans. R. Soc. London, Ser. B*, 1998, **353**, 41.

[20] T. D. Jickells, *Mar. Chem.*, 1995, **48**, 199.

[21] J. H. Martin, *Paleoceanography*, 1990, **5**, 1.

[22] T. Takahashi, R. A. Feely, R. F. Weiss, R. H. Wanninkhof, D. W. Chipman, S. C. Sutherland and T. T. Takahashi, *Proc. Natl. Acad. Sci. USA*, 1997, **94**, 8292.

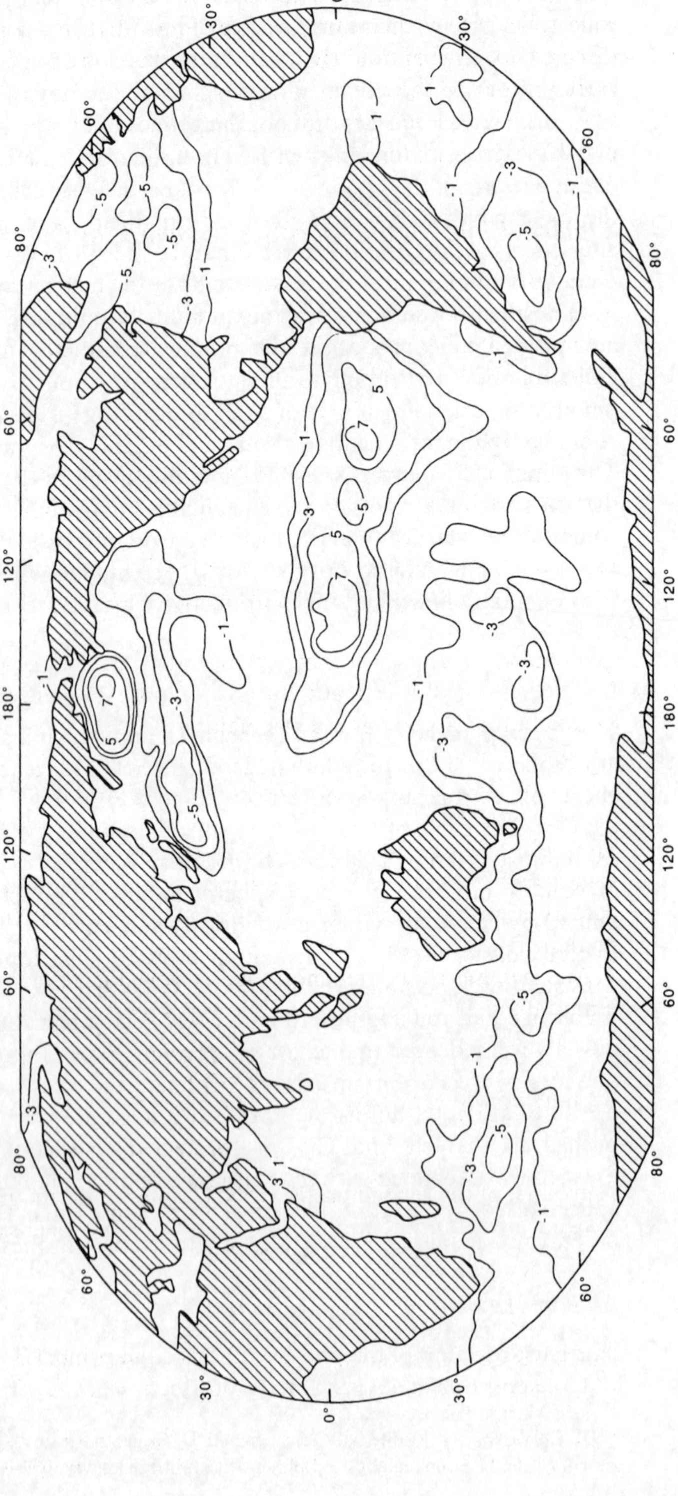

Figure 4 Mean annual net CO_2 flux over the global oceans (in 10^{12} grams of C per year per 5° square)[22]

Coastal regions are much greater sources of primary production than deeper waters, because of the nearby source of terrestrial nutrients, which enter the sea through both the air and rivers. It is also worth noting the strong seasonality of carbon input to the oceans in regions dominated by the biological control, as most such areas will have strongly enhanced carbon draw-down only during periods of maximal primary production.

Marine Biology and Oceanic Greenhouse Gas Emissions

A number of biological processes result in the marine production of gases that have a greenhouse role, similar to water vapour and CO_2. In low oxygen environments, of the sort discussed in the next section, methane is produced by anaerobic bacterial decay:

$$2(CH_2O)_n \xrightarrow{\text{bacteria}} nCH_4 + nCO_2 \tag{8}$$

The methane that escapes to the atmosphere, 1–2% of the global budget, largely derives from the sub-surface oxygen minimum associated with high productivity. The Arabian Sea in summer is the best known of such environments. Another gas produced through anaerobic decay is H_2S. This can undergo oxidation in the air to sulfate aerosols, but relatively little is likely to escape from the ocean because of its high reactivity.

Incomplete respiration, that is respiration occurring where oxygen is limited but not entirely absent and so CO_2 cannot be generated, also leads to the production of a greenhouse gas, namely CO:

$$(CH_2O)_n + \frac{n}{2}O_2 \rightarrow nCO + nH_2O \tag{9}$$

With the addition of CO caused by photochemical oxidation of methane, a significant flux enters the atmosphere annually, but the principal global contributions are terrestrial, anthropogenic and from atmospheric photolysis of methane.[23]

Another important greenhouse gas is a product of phytoplankton oxidation of ammonium. This reaction can follow two pathways:

$$2O_2 + NH_4^+ \rightarrow NO_3^- + H_2O + 2H^+ \tag{10}$$

or, much more rarely (0.1% of the time):

$$2O_2 + 2NH_4^+ \rightarrow N_2O + 3H_2O + 2H^+ \tag{11}$$

A product of the second pathway, N_2O, is a significant greenhouse gas. About a quarter of the annual input of N_2O to the atmosphere comes from the ocean.

Ocean Anoxia

The back reaction of the photosynthesis equilibrium (7)—respiration—involves the consumption of oxygen and the production of CO_2. These dissolved gases are

[23] R.T. Watson, H. Rodhe, H. Oeschger and U. Siegenthaler, in *Climate Change 1990*, ed. J.T. Houghton, G.J. Jenkins and J.J. Ephraums, Cambridge University Press, Cambridge, 1996, ch. 1, p. 1.

in the surrounding water, not within the phytoplankton cell. Thus while photosynthesis both sequesters carbon and produces oxygen, the presence of organisms at depths too deep for photosynthesis results in a depletion of the oxygen levels in this water. This depletion can also occur higher in the water column if some, perhaps dynamical, mechanism concentrates plankton at a depth where only limited photosynthesis is possible; that is, below the compensation depth.

Respirating organisms exist at all depths of the ocean, including the bottom sediments, so below the euphotic zone the ocean's supply of oxygen is slowly depleted. At the surface, oxygen levels are always around saturation. The gradual respirated depletion of oxygen as waters move away from the surface has been one way to infer the spread of water deriving from the North Atlantic throughout much of the deep waters of the global ocean.

If the replenishment of deep waters is slow, either through a particular basin being isolated from a source of deep water formation or because the oxygen utilization is faster than the re-supply of oxygenated water, then the deep water can become anoxic. Such waters will then need to use sulfates rather than oxygen in oxidizing reactions. In some regions the presence of anoxic deep water is an important climatic signal. For instance, periodically the deep eastern Mediterranean has been anoxic, as is recorded in bottom sediment layers. This is thought to be a result of enhanced surface run-off from the surrounding land masses reducing the density of the surface waters, and so preventing winter cooling from raising surface densities to a level comparable with intermediate or bottom waters and thus causing deep convection.[24] The periodicity is probably related to the Earth's orbit, through the monsoonal rainfall variation associated with mid-latitude insolation variation caused by the 20 000 year periodicity in the Earth's obliquity.

Anoxic deep water can help cause climatic change, as well as be a sign of it. Large volumes of the deep ocean can be removed as potential storage zones for carbon, because of the cessation of regional deep water formation or because an ocean basin becomes isolated from global sources of deep water renewal. If less carbon can be stored in the ocean, then more will remain in, or re-enter, the atmosphere. The atmospheric greenhouse effect will then be enhanced and the global temperature will rise. The Mediterranean is too small a basin for its periodic anoxia to cause a major direct climatic change, although changes to the exchange of water with the Atlantic caused by anoxically driven circulation changes within the Mediterranean could have indirect climatic effects.[25] A long-term cessation of North Atlantic Deep Water formation, however, or a tectonically induced isolation of a large ocean basin, could have a climatically direct effect on atmospheric CO_2. During the formation of the Atlantic Ocean, several tropical basins remained isolated from the rest of the global ocean for many millions of years and so this may have partially been responsible for the high atmospheric CO_2 levels during the Cretaceous period.[26]

[24] E.J. Rohling, *Mar. Geol.*, 1994, **122**, 1.

[25] R.G. Johnson, *Earth and Planet. Sci. Lett.*, 1997, **148**, 367.

[26] J.E. Andrews, S.K. Tandon and P.F. Dennis, *J. Geol. Soc.*, 1995, **152**, 1.

3 Oceanic Gases and Cloud Physics

Following the fate of a number of oceanically produced gases in the atmosphere reveals one of the major ways by which the ocean chemically contributes to the climate. Sulfates, sulfides and nitrogen oxides released by the ocean act as condensation nuclei within clouds, either in the form in which they were emitted from the ocean, or after undergoing chemical transitions. Sea salt, derived from evaporation of water injected into the marine boundary layer, is also an important source of cloud condensation nuclei (CCN). A greater abundance of CCN means that cloud formation, and hence precipitation, is easier to initiate, with clear implications for the surface climate. However, clouds are also major reflectors of solar radiation, and absorbers of terrestrial infrared radiation. They therefore play a fundamental role in the radiation balance of the planet. There are many indirect and direct means by which clouds affect the radiation budget,[27] some of which we will discuss in Section 4, but in general more cloud tends to result in less net energy entering the climate system.

Breaking Waves and Sea Salt

In Section 1 we discussed the basic gas exchange mechanism and saw that a transition region existed where release into the atmosphere was enhanced above some critical wind speed (Figure 2). Breaking waves, physically injecting water and its dissolved constituents into the atmosphere, cause this enhancement. In addition to enhancing the upward flux of gases, this mechanism also effectively injects sea salt into the air, through evaporation of the dispersed water droplets before they return to the sea surface.

Sea salt particles are the biggest contributor by mass of particulate material into the marine atmosphere, with 10^9–10^{10} tonnes cycled through the atmosphere annually. They tend to be relatively large, typically 0.1–1 μm in diameter. This large size means that there is a significant fall-out of particles within the marine boundary layer (up to 90%). However, those that are carried by turbulence into the free atmosphere in concentrations of typically several $\mu g\, m^{-3}$ are larger than average CCN and so play a dominant role in the important coalescence mode of rain formation.[28] This role is enhanced by the strongly hygroscopic nature of the largely NaCl sea salt aerosol. A relative humidity of only 75% is required for the initiation of condensation around a NaCl nucleus. While some salts have even lower thresholds—K_2CO_3 at 44%, for instance—the abundance of atmospheric sea salt makes this a significant source of cloud droplets.

Production Mechanisms for CCN Derived from Marine Gas Emission

Biological decay mechanisms are responsible for the emission of gases that are

[27] R. E. Dickinson, V. Meleshko, D. Randall, E. Sarachik, P. Silva-Dias and A. Slingo, in *Climate Change 1995*, ed. J. T. Houghton, L. G. Meira Filho, B. A. Callander, N. Harris, A. Kattenberg and K. Maskell, Cambridge University Press, Cambridge, 1996, ch. 4, p. 193.

[28] G. R. Bigg, *The Oceans and Climate*, Cambridge University Press, Cambridge, 1996, ch. 3, p. 85.

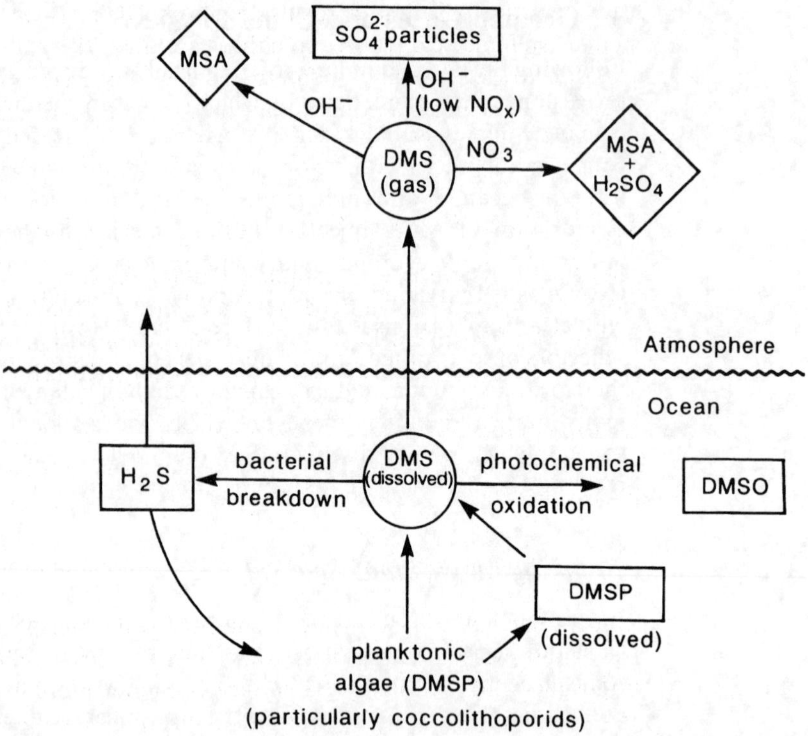

Figure 5 Schematic illustration of the sources and sinks of DMS in the marine boundary layer of the atmosphere and the oceanic mixed layer (Taken from Bigg,[28] with permission of Cambridge University Press)

precursors to CCN as well as for potential greenhouse gases. The decay of algal cells releases methyl iodide, a precursor of methyl chloride that is a natural source of chlorine, and hence ultimately involved in the chemistry of the stratosphere's ozone layer. Both dimethyl sulfide (DMS) and dimethyl sulfoniopropionate (DMSP) are produced during the destruction of the cell walls of phytoplankton, either through death, viral attack or grazing, although the exact mechanism is under constant review.[29] The production of DMSP, thought to be of importance in limiting osmotic loss of algal cell material to the surrounding seawater, varies significantly from one species to another. DMSP also oxidizes to form DMS. DMS has been observed in the marine boundary layer in significant concentrations of up to $0.25\ \mu g\ m^{-3}$ during plankton blooms.

Not all DMS released reaches the atmosphere. As can be seen from the summarizing Figure 5, some undergoes photolytic oxidation[30] within the sea through the reaction

$$2(CH_3)_2S + O_2 \xrightarrow{h\nu} 2(CH_3)_2SO \qquad (12)$$

A large proportion (30–90% in tropical waters) is absorbed by bacteria and oxidized to H_2S in order to allow the sulfur to be used by these organisms. Once in the atmosphere, DMS is oxidized by various free radicals such as hydroxyl and nitrate ions. In the presence of low concentrations of NO_x the hydroxyl reaction

[29] P.S. Liss, A.D. Hatton, G. Malin, P.D. Nightingale and S.M. Turner, *Philos. Trans. R. Soc. London, Ser. B*, 1997, **352**, 159.

[30] P. Brimblecombe and D. Shooter, *Mar. Chem.*, 1986, **19**, 343.

leads directly to sulfate aerosols, otherwise there is an acidic intermediate, methanesulfonic acid (MSA). In either case the end result is an increase in the atmosphere's acidity and sulfate-based CCN concentration, and thus cloud potential. The climatic feedback that could result from this natural marine sulfate emission is discussed in Section 4.

4 Feedback Processes Involving Marine Chemistry and Climate

The climate system is very complex. Untangling how it works involves much more than merely following the first-order energy fluxes between compartments, such as the atmosphere, Earth's surface and ocean, within the system. Feedbacks are a key characteristic leading to difficulty in predicting how perturbation of one part of the system will affect the whole. Feedbacks may be positive or negative, or occasionally self-cancelling. Often the clear first-order feedback does not reveal all the linked processes impacting on the particular feedback mechanism under discussion. This may mean that the net effect of both direct and indirect links opposes the first-order direct impact.

A major physically based feedback within the climate system is the ice albedo feedback. Ice is highly reflective, so there is a strong inverse link between global temperature and ice cover. The current increase in atmospheric concentrations of greenhouse gases does not just trap more terrestrial radiation energy in the troposphere, but warming the surface promotes more evaporation and so a greater atmospheric concentration of the dominant greenhouse gas, water vapour. However, among other effects, more clouds also result, warming affects biological activity in the ocean, and gas exchange between the atmosphere and ocean will also change. We will consider some of these inter-linked, marine-related feedbacks in more detail below.

First, however, we will consider another marine-related feedback potentially implicated in the speed of deglaciation, a theme that will also appear in the next section. This is the potential sea level and methane feedback. During glacial periods the atmospheric concentration of methane has been significantly lower than the current interglacial's pre-industrial level. This is principally because so much of the production occurs in anaerobic decay within sub-polar wetlands. These are of lesser extent during glaciations owing to the greater spread of permafrost. However, during deglaciation the rise of sea level leads to the flooding of extensive areas of permafrost on the continental shelves of the Arctic, eastern Canada and western Europe. The rapid thawing of the permafrost may lead to a corresponding release of methane trapped within the ground since the previous interglacial. As methane is a significant greenhouse gas, major, and rapid, changes in its concentration could aid climatic warming. There is some indication of such a link in Greenland ice cores,[31] but recent work has tended to weaken the likely impact of such a mechanism.[32]

[31] J. Chappellaz, T. Blunier, D. Raynaud, J. M. Barnola, J. Schwander and B. Stauffer, *Nature*, 1993, **366**, 443.

[32] R. B. Thorpe, K. S. Law, S. Bekki, J. A. Pyle and E. G. Nisbet, *J. Geophys. Res.*, 1996, **101**, 28627.

Feedbacks within the Marine Segment of the Carbon Cycle

Carbon is an important part of both the physical chemistry and the biology of the ocean. Feedbacks between the climate system and marine organisms are thought to cool the globe by about 1.7 °C today.[19] Changing the atmospheric concentration of CO_2 will therefore lead to climatic feedbacks involving the marine environment. During glacial periods a drop in atmospheric concentrations of CO_2 by about a third accompanied the methane decrease already mentioned. Such a decrease in the second most important greenhouse gas, for thousands of years, will have reduced atmospheric absorption of terrestrial radiation, and so assisted glacial cooling. Most theories to explain this decline invoke the sequestration of carbon within the marine carbon cycle. The cooling itself would promote greater oceanic absorption, through the temperature dependence of solubility and the conversion of dissolved CO_2 to bicarbonate. This is a positive feedback. Nevertheless, the enhancement of carbon draw-down associated with increased marine biological activity is needed to reconcile the magnitude of the atmospheric decline.

There are various ways by which this may have been achieved and probably more than one mechanism is required to explain the total change.[19] Increase in the supply of nutrients to the ocean is one possibility. This can be done in various ways. Atmospheric dust loadings were higher during the last glacial period. This is likely to have meant additional supplies of airborne iron being deposited in the ocean, particularly in the Southern Ocean, where winds may have been stronger because of the greater meridional temperature gradient. Greater erosion of the exposed continental shelves could have deposited more nutrients in coastal waters. Changes in ocean circulation are likely to have expanded the cold, but ice-free, regions of the North Atlantic, Southern Ocean and North Pacific. These areas are currently the strongest marine carbon sinks and so may have played an even more important role in glacial periods.

Anthropogenic change to today's atmospheric CO_2 levels is also likely to have climatic feedbacks with the ocean.[33,34] These are summarized in Figure 6. Surface warming will decrease CO_2 solubility—a positive feedback—but enhance some species' photosynthesis through the temperature dependence of the maximal growth rate—a negative feedback through the promotion of CO_2 draw-down. Radical change in species distributions opens the possibility of negating much of today's biological carbon extraction through shift of the organic carbon: $CaCO_3$ sinking ratio—a positive feedback.[34] Increased warming and greater precipitation, both predicted to be climatic changes in the 21st century, will stabilize the upper ocean water column through reducing the surface density. This, in turn, would lead to less convection transporting nutrients from below the mixed layer into the euphotic zone—a positive feedback through reducing productivity. In some, sub-tropical, regions there may be enhanced evaporation, brought about by warming and increased wind speeds, which will act in the opposite fashion—a negative feedback! Evaporation due to stronger

[33] U. Siegenthaler and J. L. Sarmiento, *Nature*, 1993, **365**, 119.

[34] K. Denman, E. Hofmann and H. Marchant, in *Climate Change 1995*, ed. J. T. Houghton, L. G. Meira Filho, B. A. Callander, N. Harris, A. Kattenberg and K. Maskell, Cambridge University Press, Cambridge, 1996, ch. 10, p. 483.

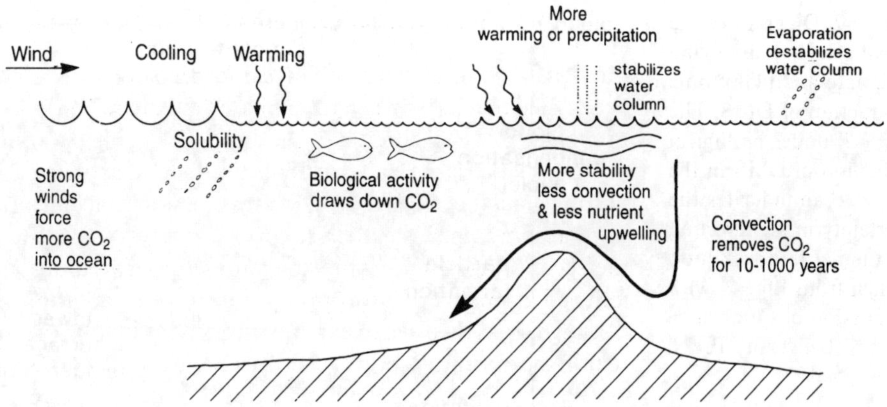

Figure 6 Schematic illustration of the mechanisms affecting absorption of CO_2 in the ocean (Taken from Bigg,[28] with permission of Cambridge University Press)

winds would also cool the sea surface, through latent heat transfer, hence both destabilizing the surface and enhancing CO_2 solubility—both negative feedbacks. A final positive feedback is via the enhanced greenhouse effect's predicted reduction of the meridional overturning circulation within the Atlantic. This currently removes large amounts of carbon from high latitude oceans, allowing biological and physicochemical processes to remove further CO_2 from the atmosphere.

The Marine Sulfur Cycle and the Charlson Hypothesis

DMS has been observed in the marine atmosphere since the early 1970s, but it was not until the mid-1980s that there was interest in this gas as being a natural source for sulfate CCN. Sulfate aerosols are, in number terms, the dominant source of CCN. The major role clouds play in the climate system leads to possible climatic implications if changes to DMS production occurred. Furthermore, the dependence of this production on environment conditions means that scope for a feedback process arises; this feedback is called the Charlson hypothesis.[35]

This feedback is illustrated in Figure 7. An increase in DMS production within the ocean leads, through oxidation of the emitted gas, to an increase in atmospheric sulfate aerosols. This, in turn, means that a greater concentration of CCN is available for any given cloud forming. Clouds will therefore have more mid-size droplets—sea salt particles provide the majority of larger droplets. Indeed, for a given cloud liquid water content, more condensation sites results, on average, in smaller droplets being produced. As the cloud's albedo depends on the surface area exposed to the incident solar beam, and the production of more small droplets tends to increase the net surface area of water droplets within the cloud, less solar radiation will penetrate the cloud. Taken to global scale, this leads to a lowering of the global temperature. The potential feedback mechanism now appears. Cooling will both lower surface ocean temperatures, thus changing the geographical distribution of where each plankton species may attain their maximal growth rate, and reduce the solar flux required for photosynthesis.

[35] R. J. Charlson, J. E. Lovelock, M. O. Andreae and S. G. Warren, *Nature*, 1987, **326**, 655.

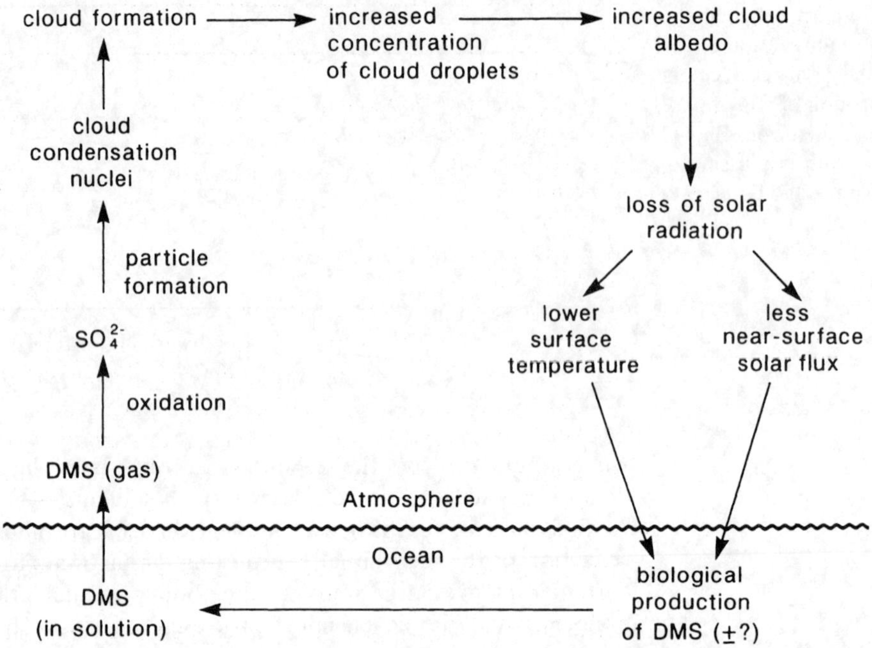

Figure 7 Diagram of the feedback loop involving climate and planktonic production of DMS. The (±?) under biological production of DMS in the ocean indicates the uncertainty in the direction of the net feedback loop (Taken from Bigg,[28] with permission of Cambridge University Press)

These changes will alter the rate of oceanic DMS production, but in a direction not yet determined. One could argue that a negative feedback would arise owing to restrictions in photosynthesis, and therefore plankton populations. However, it is also possible that equatorward penetration of the productive, cooler-water species would enhance global DMS production, giving rise to a positive feedback. It is relevant here to note that the high DMS production in the equatorial Pacific Ocean, one of the climatically most variable places in the world owing to the Southern Oscillation, remained relatively constant over the period 1982 to 1996. Some of the most dramatic, and prolonged, El Niño events of this century occurred during this interval. In the tropical Pacific, therefore, climate and marine biological productivity may be self-stabilizing to a degree through the above two opposite feedbacks.[36]

An estimate has been made of the purely direct global cooling effect due to DMS's natural enhancement of cloud albedo that is possible in today's environment.[19] This suggests that marine DMS production may be currently cooling the globe by 3.8 °C (roughly double the estimated impact of today's marine biological moderation of CO_2), although error bars on this number are large, and it does not take into account the indirect feedback mechanisms.

5 Future Prospects

The sensitivity of the ocean's chemical role in climate is yet to be fully explored as development of models for the carbon and biological components of oceanic

[36] T. S. Bates and P. K. Quinn, *Geophys. Res. Lett.*, 1997, **24**, 861

Table 1 Average annual budget of CO_2 perturbations for 1980–1989.[13] Fluxes are expressed in Gt C yr^{-1}; error limits correspond to a 90% confidence interval

	1995 estimate
CO_2 sources	
Fossil fuel emissions	5.5 ± 0.5
Tropical land-use change	1.6 ± 1.0
Reservoir change	
Atmosphere	3.3 ± 0.2
Ocean	2.0 ± 0.8
Forest re-growth	0.5 ± 0.5
Other sinks	1.3 ± 1.5

climate lag behind the physical modelling. The latter itself still has significant scope for improvement. Current estimates of the perturbation of the carbon cycle owing to anthropogenic effects also have uncertainties. Table 1 shows annual average estimates for the partitioning of carbon fluxes during the decade 1980–1989.[13] Without the ocean's enhanced uptake, atmospheric CO_2 growth would be considerably greater. Note, however, that a significant amount (1.3 Gt C yr^{-1}) of the anthropogenically emitted carbon is not clearly withdrawn from the atmosphere by a distinct mechanism. While there are a number of potential terrestrial processes that can partially account for this, for example fertilization due to higher CO_2 levels and additional nitrogen deposition, the ocean could be responsible for more carbon withdrawal than current estimates allow. Biological uptake is the largest uncertainty here. While more terrestrial sinks are being identified,[37] uncertainty still remains as oceanic estimates also continue to be revised upwards.[10,38]

Future climatic change due to increases in greenhouse gases will further alter the ocean's chemistry, and so climatic effect. Land clearance in the tropics and sub-tropics is likely to increase the dust burden in the atmosphere, and thus may lead to enhanced iron fertilization of the present High Nitrate but Low Chlorophyll (HNLC) regions of the ocean. The African Sahel appears to be already undergoing an increase in dust production due to long-term dessication.[39] There are other, atmospheric, mechanisms that result in an increased atmospheric dust load which leads to cooling, but our arguments above show that there may be a marine biologically induced cooling as well.

A number of current coupled ocean–atmosphere climate models predict that the overturning of the North Atlantic may decrease somewhat under a future warmer climate.[40] While this is not a feature that coupled models deal with well, its direct impact on the ocean's sequestration of carbon would be to cause a significant decline in the carbon that is stored in the deep water. This is a positive feedback, as oceanic carbon uptake would decline. However, the expansion of area populated by the productive cool water plankton, and the associated decline

[37] C. D. Keeling, J. F. S. Chin and T. P. Whorf, *Nature*, 1996, **382**, 146.

[38] R. L. Langenfelds, R. J. Francey, L. P. Steele, M. Battle, R. F. Keeling and W. F. Budd, *Geophys. Res. Lett.*, 1999, **26**, 1897.

[39] M. Hulme, *Geophys. Res. Lett.*, 1996, **23**, 61.

[40] A. Kattenberg, F. Giorgi, H. Grassl, G. A. Meehl, J. F. B. Mitchell, R. J. Stouffer, T. Tokioka, A. J. Weaver and T. M. L. Wigley, in *Climate Change 1995*, ed. J. T. Houghton, L. G. Meira Filho, B. A. Callander, N. Harris, A. Kattenberg and K. Maskell, Cambridge University Press, Cambridge, 1996, ch. 6, p. 285.

in the extent of sub-polar HNLC zones, may well off-set this feedback, or certainly diminish its size.[41] The greatest impact is likely to happen if atmospheric CO_2 levels reach more than double present levels, as then the North Atlantic's thermohaline circulation is likely to cease almost completely. In that case, recent model simulations suggest that there would be much less biological modification of the solubility-led uptake decrease due to surface warming.[42] This magnifies the positive physical feedback.

Changes in surface temperature elsewhere in the globe are likely to have a lesser impact on carbon or DMS production. For example, the warming that a doubling of atmospheric CO_2 could produce in the Southern Ocean has been modelled to lead to decreased carbon uptake, but enhanced biological productivity, due to the temperature effect on phytoplankton growth.[43] This would lead to an approximately 5% increase in DMS production and a lesser increase in CCN. There is thus a negative feedback here, but only of minor impact.

The role of oceanic physical chemistry and biochemistry in the enhanced greenhouse future is still uncertain. We have discussed the mechanisms generating a number of potential feedbacks, both positive and negative in their impact. However, new interactions are constantly being discovered in nature, and model representation of them is a rapidly evolving science. At present what we can say is that this is a young field of much intellectual and practical promise.

[41] J.L. Sarmiento and C. Lequere, *Science*, 1996, **274**, 1346.
[42] F. Joos, G.K. Plattner, T.F. Stocker, O. Marchal and A. Schmittner, *Science*, 1999, **284**, 464.
[43] A.J. Gabric, P.H. Whetton, R. Boers and G.P. Ayers, *Tellus B*, 1998, **50**, 388.

The Use of U–Th Series Radionuclides and Transient Tracers in Oceanography: an Overview

PETER W. SWARZENSKI, D. REIDE CORBETT, JOSEPH M. SMOAK, AND BRENT A. McKEE

1 Introduction

As we approach the 21st century, it is becoming increasingly evident that the ocean plays a highly complex and critical role in defining our global environment. For example, one of the most controversial issues is the complicated interplay between the ocean and climate change, as well as our own undeniable signature on such cycles. Exactly how the ocean influences short- and long-term climate change through exchange of heat, water, or CO_2 are topics that are being addressed in such large-scale oceanographic programs as WOCE (World Ocean Circulation Experiments) and JGOFS (Joint Global Ocean Flux Study). These studies benefit from the tremendous advances in our understanding of the environmental behavior of trace elements and radionuclides in seawater that have occurred just in the last 10–20 years. However, we still do not have an adequate grasp of the complexity of major oceanic cycles that will hopefully enable us to predict future trends in global climate change and sustainability. In this chapter we will focus on two of the most useful groups of chemical tracers in oceanography: the naturally occurring uranium- and thorium-series radionuclides and the artificially introduced transient tracers.

In the marine environment, the numerous radionuclides can be classified into three broad categories based on their production or origin: (1) those derived from the weathering of continental rocks, the *primordial* radionuclides, (2) those formed from cosmic radiation, the *cosmogenic* radionuclides, and (3) those artificially introduced into nature, the *anthropogenic* or *transient* radionuclides and tracers. The primordial radionuclides (*e.g.* ^{238}U, ^{232}Th, and ^{235}U) were incorporated during the Earth's nucleogenesis and have half-lives that are long enough to produce many shorter-lived daughter products, including isotopes of

Issues in Environmental Science and Technology No. 13
Chemistry in the Marine Environment
© The Royal Society of Chemistry, 2000

Table 1 Physical constants for some primordial radionuclides

Radionuclide	*Isotopic abundance* (%)	*Half-life* yr
^{40}K	0.0117	1.26×10^{9}
^{87}Rb	27.83	4.88×10^{10}
^{115}In	95.72	4.4×10^{14}
^{123}Te	0.905	1.3×10^{13}
^{138}La	0.092	1.06×10^{11}
^{174}Hf	0.162	2×10^{15}
^{187}Re	62.60	4.2×10^{10}
^{232}Th	100	1.41×10^{10}
^{235}U	0.72	7.04×10^{8}
^{238}U	99.27	4.46×10^{9}

Table 2 Major radionuclides produced in the atmosphere[1]

Radionuclide	*Half-life* (yr)	*Estimated production rate*[a] (atoms m^{-2} yr^{-1})
^{3}H	12.33	1.6×10^{11}
^{10}Be	1.6×10^{6}	1.26×10^{10}
^{39}Ar	269	4.2×10^{11}
^{14}C	5730	7×10^{11}
^{26}Al	7.2×10^{5}	4.8×10^{7}
^{35}S	0.24	4.8×10^{8}
^{36}Cl	3.01×10^{5}	5×10^{8}
^{81}Kr	2.13×10^{6}	—

[a]From Draganic[2]

radium, radon, polonium, bismuth, and lead. A list of the primordial radionuclides is given in Table 1. We will, however, concentrate on ^{238}U, ^{235}U, ^{232}Th, and a few of their associated daughter products that collectively make up the U–Th decay series. The uranium (^{238}U) decay series includes important isotopes of radium, radon, polonium, and lead. The uranium decay series begins with ^{238}U (half-life, $t_{1/2} = 4.46 \times 10^9$ yr) and ends with stable ^{206}Pb, after emission of eight alpha (α) and six beta (β) particles. The thorium decay series begins with ^{232}Th ($t_{1/2} = 1.41 \times 10^{10}$ yr) and ends with stable ^{208}Pb, after emission of six alpha and four beta particles. Two isotopes of radium and ^{228}Th are important tracer isotopes in the thorium decay chain. The actinium decay series begins with ^{235}U ($t_{1/2} = 7.04 \times 10^8$ yr) and ends with stable ^{207}Pb after emission of seven alpha and four beta particles. The actinium decay series includes important isotopes of actinium and protactinium. These primordial radionuclides, as products of continental weathering, enter the ocean primarily by the discharge of rivers. However, as we shall see, there are notable exceptions to this generality.

Cosmogenic radionuclides are formed in the upper atmosphere by the interaction of cosmic rays, primarily from the sun, with elements present in the atmosphere (*e.g.* ^{14}N, ^{16}O, and ^{40}Ar). Their half-lives range from months to

[1] R. M. Key, R. F. Stallard, W. S. Moore and J. L. Sarmiento, *J. Geophys. Res.*, 1985, **90**, 6995.

[2] I. V. Draganic, Z. D. Draganic and J. Adloff, *Radiation and Radioactivity on Earth and Beyond*, 2nd edn., CRC Press, Boca Raton, FL, 1993.

millions of years (Table 2) and their variable residence times in the atmosphere are controlled by physicochemical processes. These nuclides enter the ocean through wet/dry precipitation and, therefore, have a reasonably predictable geographic distribution. Cosmogenic radionuclides are generally valuable tracers for broad-scale marine processes, such as ocean circulation and water mass residence time calculations.

Finally, anthropogenic radionuclides and tracers are introduced into the environment either through nuclear reactions or industrial activity. Man-made radionuclides are usually introduced into the environment by the fission of ^{235}U and/or the detonation of fission and fusion devices below ground or in the atmosphere. These radionuclides [*e.g.* tritium (^{3}H) and ^{137}Cs] mark a specific time event or series of events that can show up in a particular horizon within a water column or a geologic record. Non-radioactive transient tracers are present in the environment as a result of a specific, quantifiable input, such as the release of chlorofluorocarbons (CFCs) to the atmosphere. Many of these tracers are derived from refrigerant and aerosol propellant use and are thus more heavily produced in the northern hemisphere. Man-made radioactive and transient tracers provide a powerful suite of tools for gaining insight into varied environmental processes. Both classes of tracers have unique attributes that make them invaluable as geochronological tools that embrace a wide range of timescales and rates of oceanic processes.

2 Radioactive Decay

The abundance of a trace element is often too small to be accurately quantified using conventional analytical methods such as ion chromatography or mass spectrometry. It is possible, however, to precisely determine very low concentrations of a constituent by measuring its radioactive decay properties. In order to understand how U–Th series radionuclides can provide such low-level tracer information, a brief review of the basic principles of radioactive decay and the application of these radionuclides as geochronological tools is useful.[3,4] The U–Th decay series together consist of 36 radionuclides that are isotopes (same atomic number, Z, different atomic mass, M) of 10 distinct elements (Figure 1). Some of these are very short-lived ($t_{1/2} \ll 1$ min) and are thus not directly useful as marine tracers. It is the other radioisotopes with half-lives greater than 1 day that are most useful and are the focus of this chapter.

Radioactive decay can be expressed as the number of atoms (N) of an unstable parent that remain at any time (t) compared to an original number of atoms (N_0):

$$N = N_0 e^{-\lambda t} \qquad (1)$$

The decay constant, λ, defines the probability that a particular atom will decay within a given time ($\lambda = \ln 2/t_{1/2}$). The half-life ($t_{1/2}$) describes a time interval after which $N = N_0/2$. The observed counting rate or activity (A) is equal to λN. Another way to describe radioactive decay is in terms of the mean life (τ) of a

[3] M. Gascoyne, in *Uranium Series Disequilibrium: Applications to Environmental Problems*, ed. M. Ivanovich and R. S. Harmon, Clarendon Press, Oxford, 1982, pp. 32–55.

[4] G. Faure, *Principles of Isotope Geology*, 2nd edn., 1991, Wiley, New York.

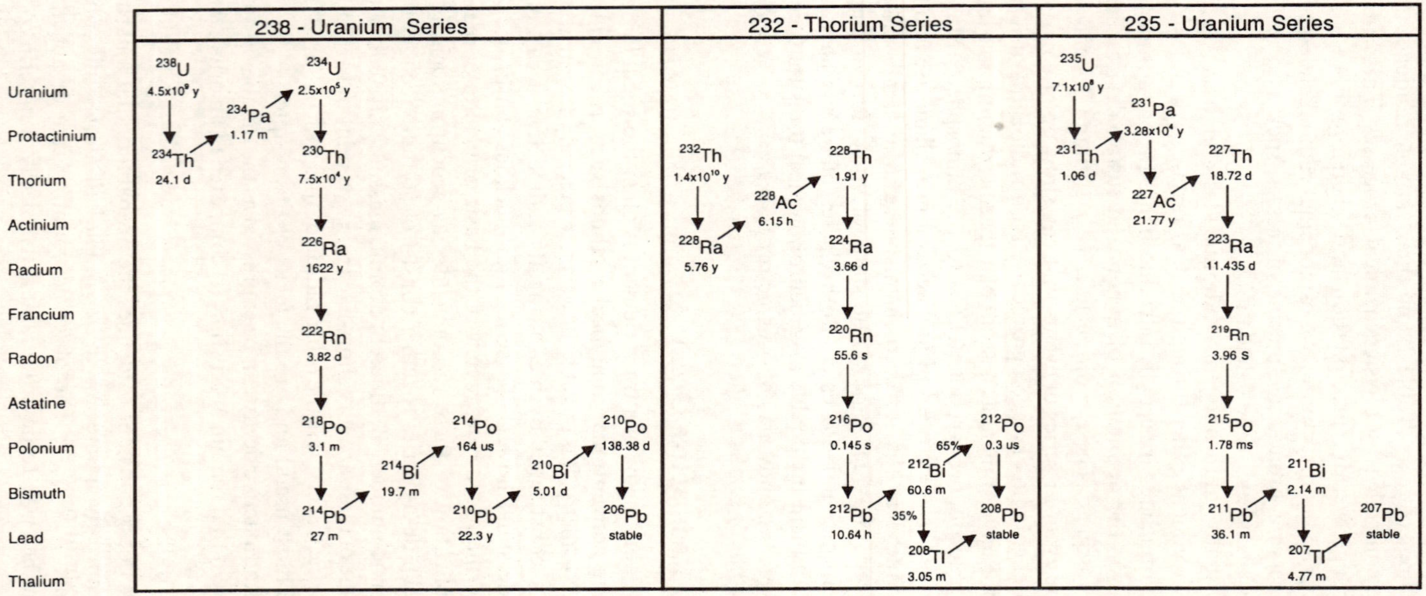

Figure 1 Chart showing the decay chain of the U–Th decay series isotopes. Vertical arrows define alpha (α) decays while beta (β) decays are illustrated by diagonal arrows

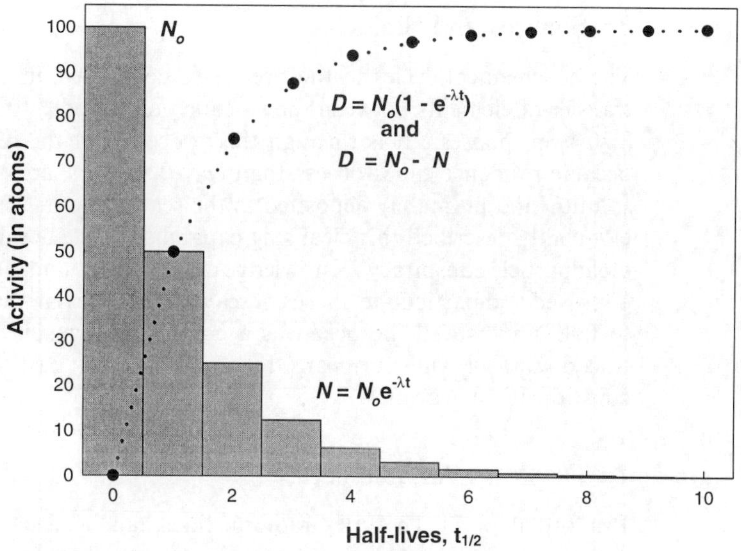

Figure 2 Exponential decay of a hypothetical radionuclide (N) to a stable daughter (D) as measured in half-lives ($t_{1/2}$). Note that as t approaches ∞, N approaches 0 and D approaches N_0 (Adapted from Faure[4])

radionuclide, which simply describes the average 'life expectancy' of a radioactive atom ($\tau = \lambda^{-1}$). Radioactive decay can thus be described either in terms of a half-life or a mean life, yet it is often more illustrative to present decay as a function of the half-life. The exponential decay of an unstable parent radionuclide (N atoms) to a stable daughter (D atoms) is illustrated in Figure 2. The number of daughter atoms produced by such decay can be expressed as:

$$D = N_0 - N \text{ and } D = N_0(1 - e^{-\lambda t}) \tag{2}$$

These are the most basic equations needed in order to utilize U–Th series radionuclides as time-dependent tracers. In turn, they can provide an 'apparent age' of a process or a geochronological framework.

An understanding of the concept of secular equilibrium is also important for such age determinations using U–Th series radionuclides. Owing to the longevity of ^{238}U, ^{232}Th, and ^{235}U, the number of parent atoms remains essentially constant for several daughter half-lives. When the rate of decay of a daughter radionuclide is equal to that of its parent, the system is considered to be at secular equilibrium (*i.e.* $\lambda_1 N_1 = \lambda_2 N_2 = \lambda_3 N_3 = \ldots$). This can be readily shown by the natural decay of ^{226}Ra ($t_{1/2} = 1622$ yr) and the ingrowth of its shorter-lived daughters (*e.g.* ^{218}Po, ^{214}Pb, ^{214}Bi, and ^{214}Po). However, many biogeochemical and physical processes, such as chemical weathering, precipitation/dissolution reactions, and inorganic/organic scavenging, can readily disrupt these decay chains. Separation of a parent radionuclide from its progeny owing to differences in chemical and physical properties is termed isotopic fractionation or radioactive disequilibrium. When this occurs, a radionuclide that has been separated from its radiogenic parent decays at a rate determined solely by its decay constant. U–Th series disequilibrium dating methods have provided oceanographers with many powerful tools to unravel both ancient (*e.g.* paleo-oceanographic) and modern (*e.g.* scavenging, ocean circulation) processes.

3 Sources and Sinks

The geochemical cycles in the ocean are controlled, in part, by the reversible transfer of elements between the so-called dissolved ($<0.4\,\mu$m) and particulate ($>0.4\,\mu$m) phases.[5] It is through this process that the reactive elements, with oceanic residence times of less than *ca.* 1000 years, are rapidly removed from solution and ultimately deposited to the seabed. Some 25 years ago, Turekian[6] eloquently described this cleansing capability of the oceans in terms of the '... great particle conspiracy'. Knowledge of the partitioning of constituents between dissolved and particulate phases in seawater is necessary for marine geochemists to fully understand the ocean as a complete biogeochemical system. Marine tracers can provide a means for tracking this partitioning or cycling of constituents in seawater.

U–Th Series Radionuclides

Primordial or U–Th series radionuclides can be discharged into the ocean through several possible pathways, illustrated schematically in Figure 3:

(a) Rivers are an important source for many radionuclides to the sea. Such transport can involve the full particle size spectrum (from particles to 'truly' dissolved). Relatively soluble species such as uranium are usually in a dissolved and/or colloidal-sized (defined loosely here as $\ll 0.4\,\mu$m) phase in fresh water. In contrast, very little dissolved thorium is present in river water. All three U–Th decay chains are represented in river water, albeit to varying degrees of fractionation or disequilibrium.
(b) Highly particle-reactive thorium isotopes settling out of the water column to the seafloor continually produce Ra–Rn isotopes. Coastal and offshore marine sediments therefore constantly release Ra and Rn into bottom waters, largely by diffusion/advection processes.
(c) Ground water (either fresh or recycled marine water) may discharge into coastal waters where hydrologic gradients or tidal forcing are favorable. It is possible that the advection of recently deposited sediments by such ground water can release radium isotopes and their progeny into coastal bottom waters.
(d) Autochthonous production of daughter isotopes by the radioactive decay of a long-lived parent can be a major oceanic source, especially in deep basins where water residence times are long.
(e) Wet/dry deposition is a final important source for cosmogenic radionuclides (and transient tracers, see next section) to the surface ocean.

Cosmogenic Radionuclides

Cosmogenic radionuclides are produced in the upper atmosphere by spallation reactions of cosmic rays with atmospheric elements. The most common

[5] W.S. Broecker and T.H. Peng, *Tracers in the Sea*, 1982, Eldigo Press, New York.
[6] K.K. Turekian, *Geochim. Cosmochim. Acta*, 1977, **41**, 1139.

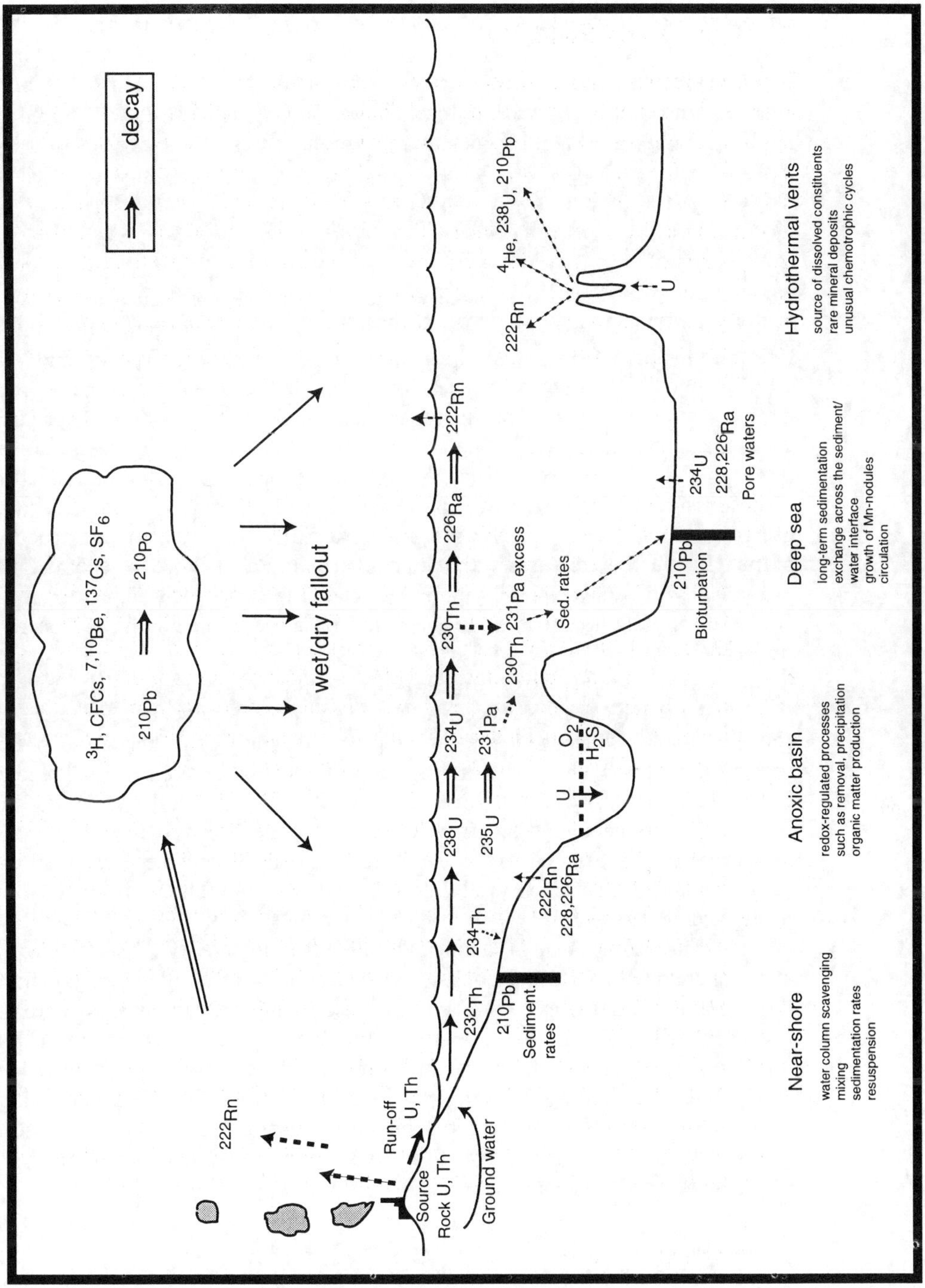

Figure 3 Schematic cartoon depicting oceanic inputs and cycles of select U–Th series isotopes

atmospherically derived nuclides used as oceanographic tracers include ^{3}H, ^{14}C, ^{7}Be, and ^{10}Be. Radioisotopes of hydrogen and carbon are of particular interest because they are major biological elements, and thus can be used as both chemical and biological tracers. Recently, particle reactive ^{7}Be ($t_{1/2}$ = 53.3 day) has been utilized as a valuable coastal tracer of short-term particle deposition and remobilization.[7–9]

Anthropogenic Tracers

Anthropogenic tracers are artificially produced, transient in nature, and may be radioactive. The chlorofluorocarbons (CFCs) or freons are organic compounds that are produced for a range of domestic and commercial purposes. CFC-11, CFC-12, and CFC-113 have long residence times in the atmosphere (on the order of 50, 180, and 90 years, respectively) and are thus evenly distributed over large geographic expanses.[10] The partial pressures of these trace gases in the atmosphere have been reconstructed from knowledge of the world's manufacture of these products. Broecker and Peng[5] noted that since freons have no place to go except the atmosphere and their rate of destruction is very low, a reconstruction of the CFC partial pressures in the atmosphere should be valid. In general, CFC partial pressures have steadily increased since the early 1960s. This feature is critical in making transient analogs effective oceanographic mixing and current tracers.[11]

Radionuclides produced during the testing of nuclear devices (*e.g.* ^{3}H, ^{90}Sr, and ^{137}Cs) may reach the sea surface within a few weeks after injection, or may also travel a great distance from the source point, depending on their atmospheric speciation and prevailing air currents. For example, the 1986 Chernobyl reactor accident released various short-lived isotopes of Kr, Sr, Zr, Ru, I, Cs, Ce, Pu, and Np into the atmosphere. While as much as 70% of the Chernobyl-released ^{137}Cs measured in Great Britain was dissolved and thus easily dispersible, a water-soluble value of only 30% was measured in Prague, even though this city is much closer to the accident site.[12]

The largest component of artificially derived radioactivity released into the atmosphere occurred during extensive nuclear weapons testing that began in the late 1940s and ended in early 1960s. Thus, these nuclides offer a pulse tracer to the oceans that can yield insight into patterns and rates of water mass circulation. It is interesting to note that the utility of several valuable atmospheric tracers (*e.g.* ^{137}Cs) is steadily diminishing, simply owing to a constituent's half-life and/or lack of a modern input. Fortunately for chemical oceanographers, the atmospheric release of other modern or longer-lived trace constituents such as ^{85}Kr and SF_6 appears to show great promise.

[7] C. R. Olsen, N. H. Cutshall and I. L. Larsen, *Mar. Chem.*, 1982, **11**, 501.
[8] E. A. Canuel, C. S. Martens and L. K. Benninger, *Geochim. Cosmochim. Acta*, 1990, **54**, 237.
[9] C. K. Sommerfield, C. A. Nittrouer and C. R. Alexander, *Cont. Shelf Res.*, 1999, **19**, 335.
[10] F. S. Rowland, *Ambio*, 1990, **19**, 281.
[11] D. W. R. Wallace, P. Beining and A. Putzka, *J. Geophys. Res.*, 1994, **99**, 7803.
[12] G. Choppin, J. O. Liljenzin and J. Rydberg, *Radiochemistry and Nuclear Chemistry*, 2nd edn., 1995, Reed Elsevier, Oxford.

Role of Particles/Surfaces

For a radionuclide to be an effective oceanic tracer, various criteria that link the tracer to a specific process or element must be met. Foremost, the environmental behavior of the tracer must closely match that of the target constituent. Particle affinity, or the scavenging capability of a radionuclide to an organic or inorganic surface site (*i.e.* distribution coefficient, K_d), is one such vital characteristic. The half-life of a tracer is another characteristic that must also coincide well with the timescale of interest. This section provides a brief review of the role of various surface sites in relation to chemical scavenging and tracer applications.

The biogeochemical processes that generally describe the interaction of elements with particles are quite well known: dissolution, flocculation, ion exchange, sorption, (co)precipitation, electron transfer, and biological uptake.[13] In aquatic environments these reactions often occur simultaneously and competitively. In order to utilize marine tracers effectively, we must understand how elements are associated with particles and sediments.

It is widely recognized that surface sites of particles are the major sequestering agent of dissolved trace elements and compounds in the aquatic environment. However, the kinetic and thermodynamic mechanisms involved in this dissolved-particulate partitioning are complex, rarely static, and generally still not well understood. Particulate material consists of various chemical components which are a result of natural chemical alterations (*e.g.* adsorption, flocculation, complexation, and redox reactions) of weathered parent material during its transport to the sea. Micro-organisms may also be important in waters that have lower concentrations of inorganic suspended material, owing to the active and/or passive incorporation of elements into soft tissues and fecal material.[14,15] Fine-grained particles and colloids generally occur as aggregates of intermeshed or layered iron oxide coatings, manganese oxide particles, organic coatings, or carbonates, and are associated predominantly with clay-sized particles.[13] These reactive surface sites are thus a source or sink for dissolved elements in the aquatic environment, depending on the chemical reactivity of a particular constituent and the water mass residence time.

Elemental particle phase associations can be used to infer chemical reactivity and oceanic behavior. The most important geochemical phase components include iron and manganese (oxy)hydroxides, organic coatings and particles, carbonates, sulfides, phosphates, and detrital clays and silicates.[13] The influence of these mineral and organic components on the chemical reactivity of a radionuclide is highly variable and related to the environmental setting. Typical master variables such as pH, temperature, dissolved oxygen, salinity, alkalinity, total CO_2, and the degree of weathering play a critical role in defining the efficiency of scavenging. For example, river-dominated coastal environments are well-known sites for colloid/particle flocculation and particle aggregation. Such areas are thus often sinks or depositional environments for many reactive

[13] W. Stumm and J. J. Morgan, *Aquatic Chemistry. Chemical Equilibria and Rates in Natural Waters*, 3rd edn., 1995, Wiley, New York.

[14] D. N. Edgington, S. A. Gordon, M. M. Thommes and L. R. Almodovar, *Limnol. Oceanogr.*, 1970, **15**, 954.

[15] D. P. Kharkar, J. Thompson, K. K. Turekian and W. O. Forstner, *Limnol. Oceanogr.*, 1976, **21**, 294.

Table 3 Historic trends in concentrations (nM) of some heavy metals in the ocean[17]

Element	*1963*[a]	*1975*[b]	*1983*[c]	*1992*[d]	*1995*[e]
Fe	180	36	0.7	0.6	0.02–1
Cu	50	8	2	2	0.5–4.5
Ag	3	0.3	0.03	0.02	0.001–0.023
Au	0.02	0.02	0.06	0.00015	0.00005–0.00015
Pb	0.2	0.2	0.005	0.005	0.003–0.150
Bi	0.1	0.1	0.05	0.00015	0.00002–0.0025

[a]From Goldberg.[18]
[b]From Brewer.[19]
[c]From Quinby-Hunt and Turekian.[20]
[d]From Nozaki.[21]
[e]From Donat and Bruland.[22]

inorganic and organic contaminants.[16] In many estuaries, one phase often does not have a very large binding capacity, and so a variety of phases may compete simultaneously, depending on local hydrological and physicochemical conditions.

4 Oceanic Behavior

The first comprehensive effort to determine the elemental composition in the world's oceans was established by the 1872–1876 British Challenger Expedition. Since then, marine geochemists have been continuously updating and expanding this dataset with highly specialized oceanographic vessels, new analytical instrumentation, contamination-free sampling protocols (Table 3), and large-scale data intercomparisons. Only within the last decade have oceanographers been able to reliably measure and agree upon trace concentrations of many dissolved constituents in seawater. These most recent developments significantly advanced our understanding of the oceanic behavior of U–Th radionuclides and enabled their use as effective tracers and geochronological tools.

Uranium

In oxygenated seawater, uranium is thermodynamically predicted to be present in a hexavalent (+6) oxidation state, but it can also exist as the tetravalent U(IV) if conditions are sufficiently reducing. Reduced uranium in the +4 oxidation state is highly insoluble or particle reactive. In contrast, U(VI) is readily soluble due to the rapid formation of stable inorganic carbonate complexes.[23] According

[16] E. R. Sholkovitz, *Geochim. Cosmochim. Acta*, 1976, **40**, 831.
[17] Y. Nozaki, in *Deep Ocean Circulation, Physical and Chemical Aspects*, ed. T. Teramoto, Elsevier, Amsterdam, 1993, pp. 83–89.
[18] E. D. Goldberg, in *The Sea*, ed. M. N. Hill, Wiley-Interscience, New York, 1963, vol. 2, pp. 3–20.
[19] P. G. Brewer, in *Chemical Oceanography*, 2nd edn., ed. J. P. Riley and G. Skirrow, Academic Press, London, 1975, vol. 1, pp. 415–496.
[20] M. S. Quinby-Hunt and K. K. Turekian, *Eos*, 1983, **64**, 130.
[21] Y. Nozaki, *Chikyukagaku*, 1992, **26**, 35.
[22] J. R. Donat and K. W. Bruland, in *Trace Elements in the Oceans*, ed. B. Salbu and E. Steinnes, CRC Press, Boca Raton, FL, 1995, pp. 247–281.
[23] D. Langmuir, *Geochim. Cosmochim. Acta*, 1978, **42**, 547.

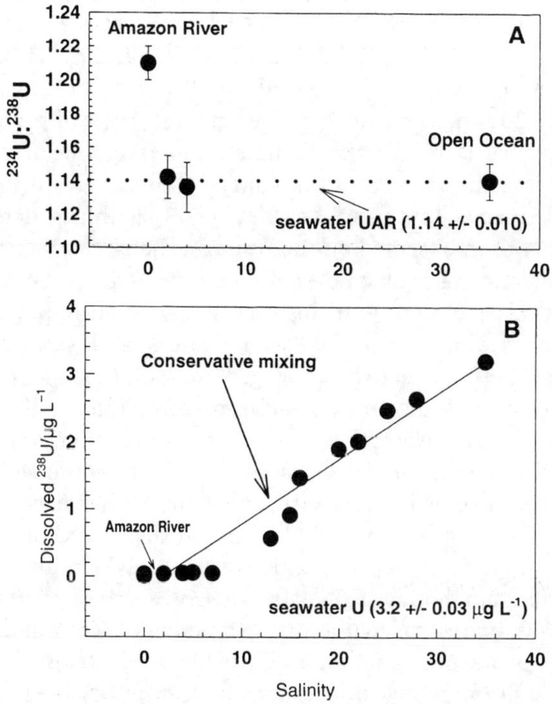

Figure 4 Measurements of (A) uranium activity ratios, UARs ($^{234}U:^{238}U$) and U concentrations (B) across a salinity gradient off the Amazon River mouth (1996). UARs were determined by thermal ionization mass spectrometry (TIMS) at Caltech (D. Porcelli); U concentrations by ICPMS

to thermodynamic calculations, in oxygenated, low pH waters, a phosphate uranium complex [$UO_2(HPO_4)_2$] might be an important species. However, at higher pHs such as in seawater, the ubiquitous carbonate complex [*e.g.* $UO_2(CO_3)_3$] prevails. Recently there has been some evidence which suggests that uranium may also exist as a peroxide complex in seawater under favorable conditions. Such a complex could potentially participate in photo-oxidation/reduction reactions, as has been shown for Mn(IV) and Fe(III).

Owing to the stability of the uranyl carbonate complex, uranium is universally present in seawater at an average concentration of *ca.* 3.2 $\mu g\,L^{-1}$ with a daughter/parent activity ratio ($^{234}U:^{238}U$) of 1.14.[24,25] In particulate matter and bottom sediments that are roughly 1×10^6 years old, the ratio should approach unity (secular equilibrium). The principal source of dissolved uranium to the ocean is from physicochemical weathering on the continents and subsequent transport by rivers. Potentially significant oceanic U sinks include anoxic basins, organic rich sediments, phosphorites and oceanic basalts, metalliferous sediments, carbonate sediments, and saltwater marshes.[26–28]

Removal of uranium from solution can take place by various mechanisms. Calcareous and siliceous organisms can efficiently remove uranium from

[24] D. L. Thurber, *J. Geophy. Res.*, 1962, **67**, 4518.
[25] M. Baskaran, P. H. Santschi, L. Guo, T. S. Bianchi and C. Lambert, *Cont. Shelf Res.*, 1996, **16**, 353.
[26] J. D. MacDougall, *Earth Planet. Sci. Lett.*, 1977, **35**, 65.
[27] F. Aumento, *Earth Planet. Sci. Lett.*, 1971, **11**, 90.
[28] T. M. Church, M. M. Sarin, M. Q. Fleisher and T. G. Ferdelman, *Geochim. Cosmochim. Acta*, 1996, **60**, 3879.

solution. Brewer[29] reported a correlation between biogenic material from sediment traps and uranium content, suggesting either active uptake by organisms or passive incorporation. However, bioaccumulation of uranium generally does not appear to be a dominant removal process in the open ocean (paucity of surface sites). Particulate uranium can be also desorbed from oxide surfaces by an increase in alkalinity or pH.[23] Desorption of uranium from suspended particles during estuarine mixing is thus geochemically feasible, and has in fact been observed on the Amazon shelf[30,31] (Figure 4). Langmuir[23] also reported a strong association between iron hydroxides and adsorbed uranium, leading to speculation that this phase may be an important sink for U.[32,33]

Marine sediments can be both sources and sinks for U. Uranium has repeatedly been shown to be scavenged from solution into reducing sediments.[34–39] This removal may also include diffusion of soluble U(VI) from seawater into the sediment via pore water. Uranium–organic matter complexes are also prevalent in the marine environment. Organically bound uranium was found to make up to 20% of the dissolved U concentration in the open ocean.[40] Uranium may also be enriched in estuarine colloids[29,31] and in suspended organic matter within the surface ocean.[38,41] Scott[42] and Maeda and Windom[43] have suggested the possibility that humic acids can efficiently scavenge uranium in low salinity regions of some estuaries. Finally, sedimentary organic matter can also efficiently complex or adsorb uranium and other radionuclides.[17,44–46]

Global uranium flux calculations have typically been based on the following two assumptions: (a) riverine-end member concentrations of dissolved uranium are relatively constant, and (b) no significant input or removal of uranium occurs in coastal environments. Other sources of uranium to the ocean may include: mantle emanations, diffusion through pore waters of deep-sea sediments, leaching of river-borne sediments by seawater,[47] and remobilization through reduction of a Fe–Mn carrier phase.[31] However, there is still considerable debate

[29] A.C. Sigleo and G.R. Helz, *Geochim. Cosmochim. Acta*, 1981, **45**, 2501.
[30] B.A. McKee, D.M. DeMaster and C.K. Nittrouer, *Geochim. Cosmochim. Acta*, 1987, **51**, 2779.
[31] P.W. Swarzenski, B.A. McKee and J.G. Booth, *Geochim. Cosmochim. Acta*, 1995, **59**, 7.
[32] R.T. Lowson, S.A. Short, B.G. Davey and D.J. Gray, *Geochim. Cosmochim. Acta*, 1986, **50**, 1697.
[33] C. Lienert, S.A. Short and H.R. von Gunten, *Geochim. Cosmochim. Acta*, 1994, **58**, 5455.
[34] C.E. Barnes and J.K. Cochran, *Earth Planet. Sci. Lett.*, 1990, **97**, 94.
[35] C.E. Barnes and J.K. Cochran, *Geochim. Cosmochim. Acta*, 1993, **57**, 555.
[36] A.M. Giblin, B.D. Batts and D.J. Swaine, *Geochim. Cosmochim. Acta*, 1981, **45**, 699.
[37] J.K. Cochran, A.E. Carey, E.R. Sholkovitz and L.D. Surprenant, *Geochim. Cosmochim. Acta*, 1986, **50**, 663.
[38] R.F. Anderson, *Geochim. Cosmochim. Acta*, 1982, **46**, 1293.
[39] R.F. Anderson, *Uranium*, 1987, **3**, 145.
[40] K. Hirose and Y. Sugimura, *J. Radiochem. Nucl. Chem. Art.*, 1991, **149**, 83.
[41] D.K. Mann and G.T.F. Wong, *Mar. Chem.*, 1993, **42**, 25.
[42] M. Scott, in *Uranium Series Disequilibria: Applications to Environmental Problems*, ed. M. Ivanovich and R.S. Harmon, Clarendon Press, Oxford, 1982, pp. 181–201.
[43] M. Maeda and H.L. Windom, *Mar. Chem.*, 1982, **11**, 427.
[44] G.N. Baturin and A.V. Kochenov, *Geokhimya*, 1973, **10**, 1529.
[45] Y. Kolodny and I.R. Kaplan, in *Proceedings of the Symposium on Hydrogeochemistry, Biogeochemistry and Hydrochemistry*, Clark, Washington, 1973, pp. 418–442.
[46] T. Mo, A.D. Suttle and W.M. Sackett, *Geochim. Cosmochim. Acta*, 1973, **37**, 35.
[47] S.G. Bhat and S. Krishnaswami, *Proc. Indian Acad. Sci.*, 1969, **70**, 1.

regarding the overall mass balance of uranium in the oceans.[34,35,48] Present estimates for the flux of uranium to the sea[49–51] include large (factor of 2) margins of uncertainty. These uncertainties result primarily from a poor understanding of: (1) the variability in riverine-end member concentrations of uranium; (2) the behavior of uranium during river/ocean mixing; and (3) the extent to which shelf sediments may influence the net flux of uranium from coastal areas.

Because of these inherent uncertainties, estimates of oceanic uranium residence times are still somewhat vague. Pioneering work on dissolved uranium in world rivers suggested that the residence time of uranium in the oceans must be on the order of 4 million years.[52] As a consequence of having such long oceanic residence times, uranium has been regarded as a relatively non-reactive (conservative) element in seawater. Subsequent studies[50,51,53,54] have broadened the data base for uranium concentrations in marine environments and current estimates of seawater uranium residence times range from 200 000 to 400 000 years. These residence time calculations are constrained by the assumption that the global riverine flux of dissolved uranium is a reasonably accurate estimate of total uranium input to the oceans.

If uranium is internally cycled in coastal environments or if the riverine delivery of U shows some variability, residence time estimates (regardless of their precision) cannot be sensitive indicators of oceanic uranium reactivity. Based on very precise measurements of dissolved uranium in the open ocean, Chen *et al.*[55] concluded that uranium may be somewhat more reactive in marine environments than previously inferred. Furthermore, recent studies in high-energy coastal environments[30,31,56,57] indicate that uranium may be actively cycled and repartitioned (non-conservative) from one phase to the next.

Thorium

Like many chemical species, thorium exhibits a great affinity for particle surfaces in the marine environment. These other species and thorium are referred to as particle reactive because they are readily removed from the dissolved phase onto the particulate phase. Thorium exists as a hydrolyzed species in seawater, $Th(OH)_n^{(4-n)+}$, and is thus extremely particle reactive. Because of its particle-reactive nature, thorium has been used to examine scavenging as an analog for other

[48] G. P. Klinkhammer and M. R. Palmer, *Geochim. Cosmochim. Acta*, 1991, **55**, 1799.

[49] J. K. Cochran, in *Uranium Series Disequilibrium: Applications to Environmental Problems*, ed. M. Ivanovich and R. S. Harmon, Clarendon Press, Oxford, 1982, pp. 384–430.

[50] W. M. Sackett, T. Mo, R. F. Spalding and M. E. Exner, in *Symposium on the Interaction of Radioactive Contaminants with the Constituents of the Marine Environment*, Seattle, WA/IAEA, Vienna, 1973, pp. 757–769.

[51] A. Mangini, C. Sonntag, G. Berstch and E. Muller, *Nature*, 1979, **278**, 337.

[52] W. S. Moore, *Earth Planet. Sci. Lett.*, 1967, **2**, 231.

[53] T. L. Ku, K. G. Knauss and G. G. Mathieu, *Deep-Sea Res.*, 1977, **24**, 1005.

[54] K. K. Turekian and L. H. Chan, in *Activation Analysis in Geochemistry and Cosmochemistry*, ed. A. O. Branfelt and E. Steinnes, Universitetsforlaget, Oslo, 1971, pp. 311–320.

[55] J. H. Chen, R. L. Edwards and G. L. Wasserburg, *Earth Planet. Sci. Lett.*, 1986, **80**, 241.

[56] J. Carroll and W. S. Moore, *Geochim. Cosmochim. Acta*, 1994, **58**, 4987.

[57] P. W. Swarzenski and B. A. McKee, *Estuaries*, 1998, **221**, 379.

particle-reactive species.[5,58] Scavenging has a major influence on the marine chemistry of many elements, including numerous metals and pollutants. Recently, thorium isotopes have been examined as proxies of biogenic flux.[59–61] Once the particle-reactive species have been scavenged, subsequent packaging and/or aggregation can result in the flux of particles and particle-reactive species from the water column. Thorium provides a unique way to study the environmental pathways and the biogeochemical processes that affect particle-reactive species. The four useful thorium isotopes are ^{232}Th ($t_{1/2} = 1.4 \times 10^{10}$ yr), ^{230}Th ($t_{1/2} = 7.5 \times 10^4$ yr), ^{228}Th ($t_{1/2} = 1.9$ yr), and ^{234}Th ($t_{1/2} = 24.1$ day).

Thorium-232 is the only non-radiogenic thorium isotope of the U/Th decay series. Thorium-232 enters the ocean by continental weathering and is mostly in the particulate form. Early measurements of ^{232}Th were by alpha-spectrometry and required large volume samples (*ca.* 1000 L). Not only did this make sample collection difficult, but the signal-to-noise ratio was often low and uncertain. With the development of a neutron activation analysis[62] and a mass spectrometry method[55,63] the quality of the data greatly improved, and the required volume for mass spectrometry was reduced to less than a liter. Surface ocean waters typically have elevated concentrations of dissolved and particulate ^{232}Th.[17,37,62] Surface water enrichment is the result of aeolian and fluvial inputs, which are thought to be the most important sources of ^{232}Th to the ocean. Thorium-232 has been proposed as a link between the radiogenic thorium isotopes and trace metals and anthropogenic pollutants.[63] While the pathways are very different for the radiogenic thorium isotopes, ^{232}Th is delivered to the ocean in a fashion similar to many pollutants and trace metals. For example, Guo *et al.*[64] found ^{232}Th distributions in the Gulf of Mexico and off Cape Hatteras in the North Atlantic Ocean agreed well with the general distribution pattern of aluminum.

Thorium-230 is supplied to seawater from *in situ* decay of its parent, ^{234}U. Both dissolved and particulate ^{230}Th activities increase with depth[65,66] in water column profiles. This increase with depth is the result of progressively more ^{230}Th being scavenged onto particles as they sink through the water column. The increase in the dissolved and particulate activity with depth indicates the scavenging of ^{230}Th can be a reversible reaction. Decay is not a factor in the ^{230}Th water column distribution owing to its long half-life. Recently ^{230}Th measurements have been coupled with ^{231}Pa to examine particle fluxes.[67,68] Protactinium-231 and ^{230}Th are supplied by ^{235}U and ^{234}U, respectively. Protactinium has a weaker affinity for particles than ^{230}Th and is therefore

[58] P. H. Santschi, Y.-H. Li and J. J. Bell, *Earth Planet. Sci. Lett.*, 1979, **45**, 201.
[59] R. W. Eppley, in *Productivity of the Ocean: Present and Past*, ed. W. H. Berger, V. S. Smetacek and G. Wefer, Wiley, New York, 1989, pp. 85–97.
[60] K. O. Buesseler, M. P. Bacon, J. K. Cochran and H. D. Livingston, *Deep-Sea Res.*, 1992, **39**, 1115.
[61] J. M. Smoak, W. S. Moore, R. C. Thunell and T. J. Shaw, *Mar. Chem.*, 1999, **65**, 177.
[62] C.-A. Huh and M. P. Bacon, *Anal. Chem.*, 1987, **57**, 2138.
[63] C.-A. Huh and T. M. Beasley, *Earth Planet. Sci. Lett.*, 1987, **85**, 1.
[64] L. M. Guo, M. Baskaran, A. Zindler and P. H. Santschi, *Earth Planet. Sci. Lett.*, 1995, **133**, 117.
[65] P. G. Brewer, Y. Nozaki, D. W. Spencer and A. P. Fleer, *J. Mar. Res.*, 1980, **38**, 703.
[66] M. P. Bacon and R. F. Anderson, *J. Geophys. Res.*, 1982, **87**, 2045.
[67] H. N. Edmonds, S. B. Moran, J. A. Hoff, J. N. Smith and R. L. Edwards, *Science*, 1998, **280**, 405.
[68] N. R. Kumar, R. Gwiazda, R. F. Anderson and P. N. Froelich, *Nature*, 1993, **362**, 45.

susceptible to lateral transport. Thorium-230 flux from the water column roughly balances its local rate of production. However, ^{231}Pa fluxes have been observed to increase with an increase in particle flux.[69] In areas of high particulate flux, ^{231}Pa:^{230}Th ratios can exceed the production ratio. This technique has been exploited in sediment cores to examine particle flux variability through time and hence productivity.[68]

The two remaining thorium isotopes have much shorter half-lives. Thorium-228 has a half life of 1.91 years and is produced from ^{228}Ra, with ^{228}Ac as a very short-lived intermediate daughter. The ^{228}Th:^{228}Ra ratio can be used to calculate residence times with respect to uptake on particles.[5] Thorium-228 is an excellent tracer for intermediate timescale processes, while ^{234}Th is an ideal tracer of short timescale processes because the half-life is 24.1 days and the only source is the decay of ^{238}U in seawater. Thorium-234 also has been used to examine residence times with respect to uptake on particles and removal of particles from the surface ocean.[70] Thorium-234 was first measured in the ocean by Bhat and Krishnaswami.[47] Thorium-234 generally is deficient with respect to its parent ^{238}U in surface waters as a result of the scavenging by particles and particle flux. Below approximately 200 m in the water column, ^{234}Th and ^{238}U are typically in secular equilibrium and the uptake of ^{234}Th is balanced by the loss from decay. In the open ocean, biogenic particles dominate the particle flux and therefore are the scavengers of thorium. In a series of papers,[71–73] the link between biogenic particles and removal of ^{234}Th was established. They showed that ^{234}Th scavenging rates varied as a function of new biomass production. Based on this work it was proposed that the flux of ^{234}Th out of the surface ocean could be used to estimate new biomass production. More recently, Buesseler *et al.*[74,75] pioneered this work by measuring the depletion of ^{234}Th in the water column to predict the flux of particulate organic carbon (POC). Buesseler's approach was to measure the ^{234}Th/POC ratio of sinking particles, then simply multiply the export rate by the ratio. This approach has been widely used during JGOFS.[76,77]

Radium

The historic discovery of radium in 1898 by Marie Curie initiated a remarkable use of this isotope as an early oceanic tracer. Less than 10 years after its discovery,

[69] Y. Lao, R. F. Anderson, W. S. Broecker, H. J. Hofmann and W. Wolfli, *Geochim. Cosmochim. Acta*, 1993, **57**, 205.

[70] M. Baskaran, P. H. Santschi, G. Benoit and B. D. Honeyman, *Geochim. Cosmochim. Acta*, 1992, **56**, 3375.

[71] K. W. Bruland and K. H. Coale, in *Dynamic Processes in the Chemistry of the Upper Ocean*, ed. J. D. Burton, P. G. Brewer and R. Chesselet, Plenum Press, New York, 1986, pp. 159–172.

[72] K. H. Coale and K. W. Bruland, *Limnol. Oceanogr.*, 1987, **32**, 189.

[73] K. H. Coale and K. W. Bruland, *Limnol. Oceanogr.*, 1985, **30**, 22.

[74] K. O. Buesseler, L. Ball, J. Andrews, C. Benitez-Nelson, R. Belastock, F. Chai and Y. Chao, *Deep-Sea Res.*, 1998, **45**, 2461.

[75] K. O. Buesseler, *Global Biogeochem. Cycles*, 1998, **12**, 297.

[76] J. W. Murray, J. Young, J. Newton, J. Dunne, T. Chapin and B. Paul, *Deep-Sea Res.*, 1996, **43**, 1095.

[77] M. M. Rutgers van der Loeff, J. Friedrich and U. V. Bathmann, *Deep-Sea Res.*, 1997, **44**, 457.

Joly[78] observed elevated ^{226}Ra activities in deep-sea sediments that he attributed to water column scavenging and removal processes. This hypothesis was later challenged with the first seawater ^{230}Th measurements (parent of ^{226}Ra), and these new results confirmed that radium was instead actively migrating across the marine sediment–water interface. This seabed source stimulated much activity to use radium as a tracer for ocean circulation. Unfortunately, the utility of Ra as a deep ocean circulation tracer never came to full fruition as biological cycling has been repeatedly shown to have a strong and unpredictable effect on the vertical distribution of this isotope.

In the U–Th decay series there are four radium isotopes, ^{223}Ra ($t_{1/2} = 11.4$ day), ^{224}Ra ($t_{1/2} = 3.6$ day), ^{228}Ra ($t_{1/2} = 5.7$ yr), and ^{226}Ra ($t_{1/2} = 1622$ yr), with half-lives that coincide well with the timescales of many coastal and oceanic processes.[79] Two important geochemical characteristics make radium ideal as a marine tracer: (1) having highly particle-reactive thorium as its direct radiogenic parent which ties it directly to bottom sediments, and (2) exhibiting vastly different environmental behavior in fresh water and saltwater systems. Both of these criteria control the production and input of radium in marine systems. In coastal waters, thorium is efficiently scavenged by particles/colloids and rapidly removed to the seabed. In fresh water, radium is bound strongly onto particle surfaces; however, as the ionic strength of a water mass increases during mixing into seawater, desorption reactions and the general absence of particles in the open ocean maintain radium entirely in a dissolved phase.[49] Estuarine sediments thus provide a continuous source for radium isotopes to coastal waters, and the production rate is defined directly by their individual isotopic decay constants.[80,81] Source functions for radium in an estuary may thus include the following components: (1) riverine, (2) oceanic, (3) estuarine sediments, and (4) groundwater.[82–84] The relative significance of each of these source terms is defined by the particular environmental constraints of an estuary.

Ground-water discharge, which can consist of recycled marine water or fresh water, may enhance the diffusive/advective flux of dissolved radium from the seabed to a coastal water column wherever the hydraulic gradients and sediment transmissivities are favorable.[84] In coastal systems where ground water is discharged to coastal waters either continuously or ephemerally owing to tidal forcing, a localized disequilibrium between ^{228}Ra, ^{228}Th, and ^{224}Ra can develop. This disequilibrium occurs because dissolved radium will be rapidly advected into the water column while thorium remains attached to bottom sediments. In surficial sediments that are flushed by the upward movement of groundwater, ^{224}Ra : ^{228}Ra activity ratios can become quite large as a result of this process and can be used to model the flux across the sediment/water interface. Below this diagenetically active surface sediment layer, one can expect secular equilibrium to

[78] J. Joly, *Philos. Mag.*, 1908, **15**, 385.
[79] P. W. Swarzenski, *U.S. Geol. Surv. Circ.*, 1998, FS-065-99.
[80] W. S. Moore and J. F. Todd, *J. Geophys. Res.*, 1993, **98**, 2233.
[81] I. A. Webster, G. J. Hancock and A. S. Murray, *Geochim. Cosmochim. Acta*, 1995, **59**, 2469.
[82] M. S. Bollinger and W. S. Moore, *Geochim. Cosmochim. Acta*, 1993, **57**, 2203.
[83] Rama and W. S. Moore, *Geochim. Cosmochim. Acta*, 1996, **60**, 4645.
[84] W. S. Moore, *Nature*, 1996, **380**, 612.

develop from ^{232}Th down to ^{224}Ra. Such radioactive equilibrium in the ^{232}Th decay series requires approximately 20 years of sediment storage. Resuspension, bioturbation, chemical dissolution as well as groundwater flow will allow these two sediment layers to interact. Moore and colleagues have used such disequilibria to derive a groundwater flux rate in a South Carolina salt marsh.[82,83] Finally, short-lived ^{224}Ra: ^{223}Ra activity ratios may also be used to derive apparent estuarine water mass ages; higher ratios generally imply younger waters owing to the more rapid decay of radium.

Radon-222

Inert ^{222}Rn is a unique marine tracer in that it should not interact in any biogeochemical cycles. It has been used extensively for determining exchange rates across both air–sea and sediment–water interfaces. In addition, ^{222}Rn has also been used as a tracer for evaluating coefficients of isopycnal and diapycnal mixing in the deep sea[5,85] and, more recently, for tracing submarine groundwater discharge into the coastal zone.[86,87] Radon is produced through the direct alpha decay of ^{226}Ra and has a relatively short half-life of 3.85 days. With such a short half-life, ^{222}Rn is typically found in equilibrium with its parent away from the air–water and sediment–water interfaces. However, near the air–water and sediment–water interfaces, disequilibria of radon with respect to its parent readily prevail.

There are two main sources of ^{222}Rn to the ocean: (1) the decay of sediment-bound ^{226}Ra and (2) decay of dissolved ^{226}Ra in a water column. Radon can enter the sediment porewater through alpha recoil during decay events. Since radon is chemically inert, it readily diffuses from bottom sediments into overlying waters. The diffusion of radon from sediments to the water column gives rise to the disequilibrium (excess ^{222}Rn) observed in near-bottom waters. Radon is also continuously being produced in the water column through the decay of dissolved or particulate ^{226}Ra.

Loss of radon in the ocean occurs typically through radioactive decay (producing four short-lived daughters before decaying to ^{210}Pb) or loss to the atmosphere at the air–sea interface. Loss of radon owing to turbulence or diffusion at the air–sea interface leads to a depletion of radon with respect to ^{226}Ra, allowing for studies on gas exchange at this interface.[5]

Lead-210 and Polonium-210

The majority of published ^{210}Pb reports address the utility of ^{210}Pb as a geochronological tool rather than as an element that is involved in complex biogeochemical cycles. Nonetheless, some of these studies do provide insight into the geochemical behavior of ^{210}Pb and ^{210}Po. Nearly all of the lead in the world's surface oceans is believed to be of anthropogenic origin—derived from combustion

[85] J. L. Sarmiento and C. G. Rooth, *J. Geophys. Res.*, 1980, **85**, 1515.

[86] J. Cable, W. Burnett, J. Chanton and G. Weatherly, *Earth Planet. Sci. Lett.*, 1996, **144**, 591.

[87] D. R. Corbett, J. Chanton, B. Burnett, K. Dillon, and C. Rutkowski, *Limnol. Oceanogr.*, 1999, **44**, 1045.

of leaded gasoline.[88] Even the most remote areas of the world—the polar caps—show a very clear effect of lead pollution. Recent measurements of polar ice cores show that man's activities have caused a 300-fold increase in Pb since the beginning of the industrial revolution.[89] Lead-210 and its indirect daughter ^{210}Po are both important radionuclides used in geochemical research. In broad terms, the marine cycle of ^{210}Pb is characterized by supply or production of dissolved ^{210}Pb which is then scavenged by solid surfaces and thus removed from the water column. Lead-210 is delivered to the water column by three sources: (1) atmospheric fallout from the decay of ^{222}Rn, (2) from the *in situ* decay of ^{226}Ra and ^{222}Rn in the water column, and (3) from direct leaching of rocks and soils during weathering.[3] Polonium-210 is a decay product of ^{210}Pb and is produced mainly by this source in the water column, but this isotope may also have an atmospheric source term within surface waters. Lead-210 has been used extensively as a tracer for particle-reactive elements as well as for geochronological studies of lake, estuarine, and coastal sediments.

Lead-210 is rapidly removed by various processes in the water column and deposited onto the seafloor during particle deposition and accumulation. The mechanisms for removal may be similar to thorium scavenging, as studies in the New York Bight indicate similar removal times for both ^{210}Pb and ^{228}Th.[90] Soil profiles have shown organic matter to be effective in binding ^{210}Pb.[91] River water is extremely depleted in dissolved ^{210}Pb and ^{234,228}Th under normal conditions. In areas where pH may be low (*e.g.* where acid mines drain into water), elevated ^{210}Pb concentrations have been reported.[92] As this acid mine water is diluted with natural waters, the sequential precipitation of iron oxides followed by manganese oxides appears to completely scavenge ^{210}Pb and other metals onto suspended particles. In the ocean, sinking particles are adsorption sites for ^{210}Pb and, even here, iron and manganese oxides are important removal phases.[93,94] In anoxic environments, ^{210}Pb may be readily removed as a lead sulfide precipitate, causing this mineral component to be of importance in sulfide-rich sediments.

Involvement in biological cycles may affect the distribution of both ^{210}Pb and ^{210}Po. Santschi *et al.*[58] noticed large seasonal variations of ^{210}Pb and ^{210}Po in Narragansett Bay. This seasonal trend was ascribed to two factors: remobilization out of the sediments during the spring and early summer, or formation of organic complexes. Polonium-210 appears to be somewhat more reactive than ^{210}Pb in seawater.[95] In a recent summary paper, Nozaki[21] suggested that the removal of ^{234}Th, ^{210}Pb, and ^{210}Po from marine waters may be accelerated by biological

[88] B. K. Schaule and C. C. Patterson, in *Trace Metals in Sea Water*, ed. C. S. Wong, E. Boyle, K. W. Bruland, J. D. Burton and E. D. Goldberg, Plenum Press, New York, 1983, pp. 487–503.

[89] A. Ng and C. C. Patterson, *Geochim. Cosmochim. Acta*, 1981, **45**, 2109.

[90] Y.-H. Li, P. H. Santschi, A. Kaufman, L. K. Benninger and H. W. Feely, *Earth Planet. Sci. Lett.*, 1981, **55**, 217.

[91] L. K. Benninger, D. M. Lewis and K. K. Turekian, in *Marine Chemistry in the Coastal Environment*, ed. T. M. Church, ACS Symposium Series, no. 18, American Chemical Society, Washington, 1975, pp. 202–210.

[92] D. M. Lewis, *Geochim. Cosmochim. Acta*, 1977, **41**, 1557.

[93] L. S. Balistrieri, J. W. Murray and P. Paul, *Geochim. Cosmochim. Acta*, 1995, **59**, 4845.

[94] G. Benoit and H. F. Hemond, *Geochim. Cosmochim. Acta*, 1987, **51**, 1445.

[95] D. Kadko, *J.Geophys. Res.*, 1993, **98**, 857.

activity. Preferential biological uptake of ^{210}Po relative to ^{210}Pb in some sediments may cause local disequilibria, which can affect the interpretation of ^{210}Pb-derived geochronologies.[94] Hodge *et al.*[96] demonstrated that Po, U, and Pu have different uptake kinetics onto solid phases. They exposed different solid phases to seawater to determine whether Po, U, or Pu could be removed by inorganic material. Their results suggest that organic surface coatings may collect particles, with their associated elements, at differing rates. Such variable uptake rates led the authors to infer that particulate Pu, Po, and U may be associated with different particulate phases.

CFCs

Since the early 1980s, chlorofluorocarbons (CFCs) have been used to trace oceanic circulation and mixing processes. These compounds are ideally suited for this purpose as they have no natural sources and the anthropogenic sources are well understood. Some CFCs can be photochemically destroyed by UV light in the stratosphere, but they are very stable in the troposphere and in the ocean.[79] CFCs have been increasing in the atmosphere since the onset of their production and are well distributed in both hemispheres. This is due to their rather long atmospheric lifetimes (>70 years) and the rather short time of atmospheric exchange between hemispheres (*ca.* 2 years). This worldwide distribution makes these trace gases ideal tracers for oceanic studies at high latitudes, where significant quantities of water are down-welled to form intermediate and deep water masses. Unlike many other anthropogenic tracers, CFCs can be measured at sea using a gas chromatograph (GC) equipped with an electron capture detector (ECD) and a purge and trap system, providing results soon after sample collection.

Two types of CFCs, CFC-11 (CCl_3F) and CFC-12 (CCl_2F_2), have been produced since 1940. Their atmospheric histories are known since 1981 (Figure 5) and have been reconstructed using estimates of production and releases obtained from the Chemical Manufacturers Association (1983). Prior to 1940, the atmosphere and ocean were CFC free. Upon atmospheric release, gas exchange began delivering CFCs to the ocean. Therefore, recently ventilated waters contain relatively high concentrations of CFCs. These CFC concentrations can be used to deduce when a water mass was last in contact with the atmosphere. Owing to mixing processes, CFC-enriched waters can become diluted with older, CFC-free waters. For this reason, the CFC-11:CFC-12 ratio is used to estimate the age of a given water mass since this ratio is unaffected by dilution (Figure 6). In some situations the CFC-enriched waters may mix with other waters containing CFCs, with the result that the ratio will reflect an intermediate age. This age is usually biased towards the younger component, which carries higher CFC concentrations.[11]

The atmospheric ratio of CFC-11:CFC-12 increased until 1975, when regulation of the USA production of these compounds led to a reduction in their rate of increase. Since this time, the ratio has remained fairly constant. This makes

[96] V. F. Hodge, M. Koide and E. D. Goldberg, *Nature*, 1978, **277**, 206.

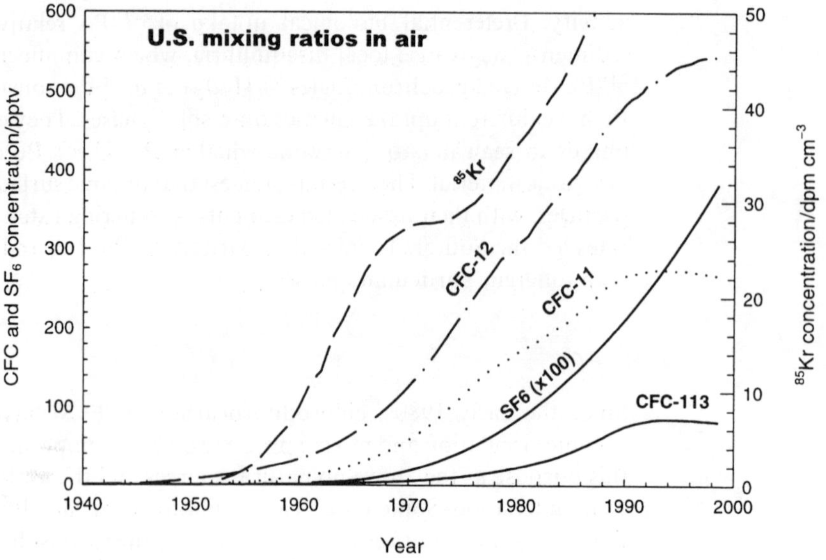

Figure 5 Atmospheric concentrations for CFC-11, CFC-12, CFC-113, SF_6 (pptv) and ^{85}Kr (dpm cm^{-3}). CFC data courtesy of E. Busenberg and L. N. Plummer (US Geological Survey); atmospheric ^{85}Kr activities complied by W. M. Smethie (LDEO) (dpm = disintegrations per minute)

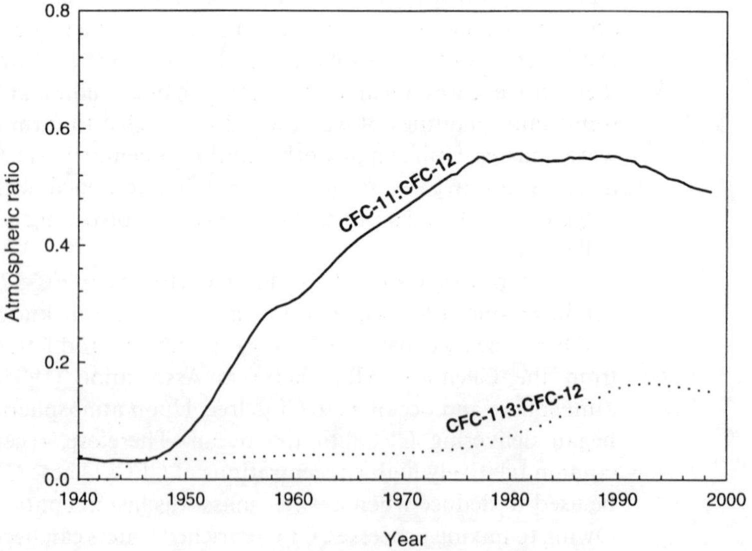

Figure 6 Atmospheric ratios of CFC-11: CFC-12 and CFC-113: CFC-12

it difficult to distinguish between water masses that have formed since this time. However, an additional CFC, CFC-113 (CCl_2FCClF_2), entered production during the 1960s. This trace gas is used as a solvent in the electronics industry.[97] Introduction of this gas into the atmosphere has led to a steadily increasing CFC-113: CFC-11 ratio. This ratio provides a much finer resolution of a water mass's age for waters ventilated since 1977. In fact, it is possible to estimate an apparent age with an error of less than a few years.

[97] D. P. Wisegarver and R. H. Gammon, *Geophys. Res. Lett.*, 1988, **15**, 188.

Many studies have shown the usefulness of CFCs as tracers of the formation and circulation of recently ventilated oceanic waters.[11,49,98–101] These studies have shown that water along the equator is relatively young (<15 years) at shallow isopycnal surfaces of North Pacific intermediate water and becomes older with depth and density (*ca.* 40 years). Intermediate waters are formed in both hemispheres and have been studied with traditional physical tracers (*i.e.* salinity and temperature) for many years. These studies show that these water masses can be traced over time and space using CFCs. This new information has generally supported previous studies of ocean circulation patterns but can resolve them better through time. However, most studies have utilized CFC-11 and CF-12, underestimating the utility of CFC-13. Perhaps future work will employ all three trace gases as transient tracers, potentially providing even more precise information.

[98] M. J. Warner and R. F. Weiss, *Deep-Sea Res.*, 1992, **39**, 2053.

[99] M. J. Warner, J. L. Bullister, D. P. Wisegarver, R. H. Gammon and R. F. Weiss, *J. Geophys. Res.*, 1996, **99**, 20 525.

[100] Y. W. Watanabe, K. Harada and K. Ishikawa, *J. Geophys. Res.*, 1994, **99**, 25 195.

[101] M. P. Meredith, K. A. Van Scoy, A. J. Watson and R. A. Locarinini, *Geophys. Res. Lett.*, 1996, **23**, 2943.

Pharmaceuticals from the Sea

RAYMOND J. ANDERSEN AND DAVID E. WILLIAMS

1 Introduction

There are a large number of life-threatening or chronically debilitating human diseases such as solid tumor cancers, AIDS, antibiotic-resistant microbial infections, asthma, and diabetes that urgently require improved medical treatments. Drug therapy represents one well-established and still attractive approach to treating these serious diseases. In order for the chemotherapeutic approach to be more effective, there is a pressing need to discover and develop new drugs that act against cancer cells, viruses, microbial pathogens, and other molecular disease targets by novel biochemical mechanisms and have diminished side effects. Secondary metabolites produced by the plants, animals, and microorganisms living in the world's oceans represent a vast and relatively unexplored resource of structurally diverse low molecular weight organic molecules that are ideal raw materials for the development of new drugs. Exploitation of this extremely valuable molecular resource is a complex and lengthy process that is shaped by many scientific, legal, business, and environmental issues. The goal of this chapter is to highlight the tremendous opportunities and the considerable challenges involved in developing new pharmaceuticals from the sea.

The majority of human drugs (see Figure 1) are relatively low molecular weight organic molecules with masses in the range of *ca.* 200 [aspirin (**1**): a pain killer] to 1500 [vancomycin (**2**): an antibiotic] Daltons.[1] Most of the drugs currently used in the clinic come from two fundamentally different origins. One large group are secondary metabolites, commonly called 'natural products', isolated from terrestrial plants or cultures of soil microorganisms.[2] Two important examples are paclitaxel (**3**), a clinically important anticancer drug that was originally isolated from the Pacific Yew tree *Taxus bevifolia*, and penicillin G (**4**), an antibiotic produced in culture by the fungus *Penicillium notatum*. In some

[1] S. Aldridge, *Magic Molecules, How Drugs Work*, Cambridge University Press, Cambridge, 1998, ch. 2, p. 23.
[2] S. Grabley and R. Thiericke (ed.), *Drug Discovery from Nature*, Springer, Berlin, 1999.

Issues in Environmental Science and Technology No. 13
Chemistry in the Marine Environment
© The Royal Society of Chemistry, 2000

Figure 1 Drugs derived from natural products

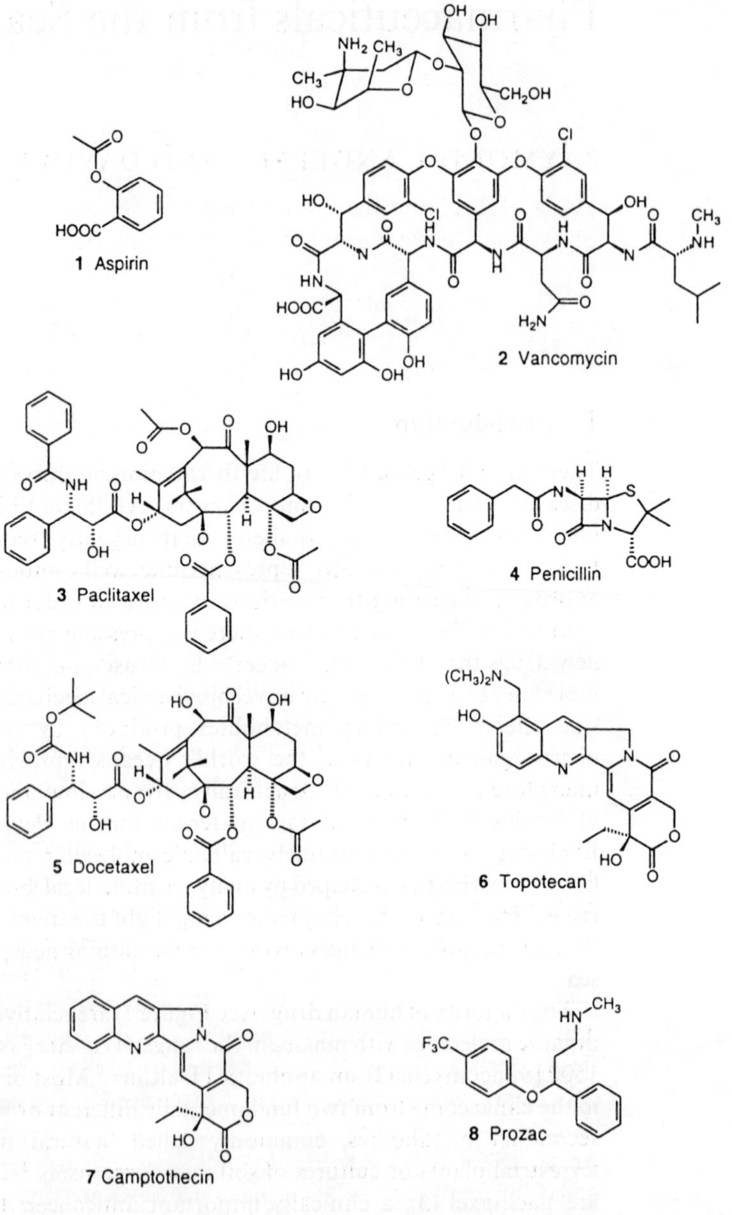

instances, the plant or microbial metabolite is subjected to synthetic modification to give a semisynthetic derivative or else it serves as a model pharmacophore to guide the synthesis of structural analogs that have enhanced drug properties. Docetaxel (**5**) is a semisynthetic analog of paclitaxel that has also been approved for clinical use as an anticancer drug and topotecan (**6**), an anticancer drug currently in clinical trials, is a synthetic analog of the plant metabolite camptothecin (**7**) isolated from the stem wood of *Camptotheca acuminata*, which itself has anticancer properties. The second group of drug molecules, which are totally synthetic in origin, are not based on any naturally occurring template. Prozac [fluoxetine (**8**)], a widely used antidepressant, is classic example of a totally synthetic drug.

Historically, drug discovery programs have relied on *in vitro* intact tissue or cell-based assays to screen libraries of synthetic compounds or natural product extracts for pharmaceutically relevant properties. This screening approach identifies compounds that elicit a desired cellular response such as antibacterial or cytotoxic activity regardless of the mechanism of action. Cell-based or intact tissue assays are relatively expensive and labor intensive, characteristics that ultimately limit the total number of assays performed and the rate at which they are executed in any drug screening program. An excellent example of a cell-based screen is the US National Cancer Institute's In Vitro Cell Line Screening Project (IVCLSP) that was operational in the early 1990s.[3] This screen, which used 60 different human tumor cell lines, was designed to identify compounds that had potential anticancer activity by looking for selective growth inhibition or cell killing of particular cell lines. IVCLSP, which at its peak screened up to 20 000 pure compounds or natural product extracts per year, illustrates the typical scale for traditional cell-based screens.

Research in molecular biology during the last two decades has provided enormous advances in our understanding of many diseases at the gene or molecular level. For example, there are now many well-documented genetic differences between cancer cells and normal cells.[4] The most commonly altered gene products in human solid tumors include ErbB2/HER (a membrane-associated tyrosine kinase), RAS (a GTP-binding protein), and proteins that affect transcription, such as Myc, Rb, or p53. These and other genetically validated protein targets have provided exciting new opportunities for developing selective mechanism-based inhibitors of specific biochemical processes that are essential to the malignant phenotype of cancer cells. They offer the hope of developing anticancer drugs with low toxicity and, therefore, much improved therapeutic margins. Similar situations exist for a whole range of other diseases.

It has been estimated that all the drugs in current clinical use target no more than 420 receptors, enzymes, ion channels, and other as yet unknown proteins.[5,6] The Human Genome Project has the potential to identify a large number of additional interesting drug targets. One conservative estimate has suggested that for the 100–150 major diseases for which improved therapies are needed there are perhaps 3000–10 000 new drug targets that will emerge from genomics. Once a molecular target and its relevance for human therapy have been identified, its large scale production for drug screening is usually carried out in a suitable genetically engineered microbial expression system. This can be relatively straightforward for some enzymes but may be considerably more difficult for receptors and other complex proteins.

In a parallel development, there have recently been tremendous advances in

[3] M. R. Boyd and K. D. Paull, *Drug Dev. Res.*, 1995, **34**, 91.
[4] J. B. Gibbs and A. Oliff, *Cell*, 1994, **78**, 193.
[5] J. Drews, *Nat. Biotechnol.*, 1996, **14**, 1516.
[6] J. Lehman *et al.*, *Nature*, 1996, **384**, suppl., 1.

the robotic technology for screening chemical entities.[2] The fusion of molecular-target-based screening and robotics has resulted in the current drug screening paradigm known as 'high-throughput screening' (HTS).[7] HTS is the process by which large numbers of compounds are tested in an automated fashion for activity as inhibitors or activators of a particular molecular target such as a cell surface receptor or metabolic enzyme. The bioassays used in HTS programs are either *in vitro* screens against purified enzymes or receptors or cell-based assays using engineered eukaryotic cells or microorganisms. HTS programs in major drug companies can typically handle 3000 enzyme inhibition or 4800 engineered cell-based assays per day, making it feasible to screen libraries of up to 100 000 or more chemical entities in a reasonable time frame.[2] The primary role of HTS is to identify 'lead' chemical structures that act against a particular molecular drug target and supply directions for their structure–activity optimization.

Key elements in successful HTS programs are the judicious choice of validated molecular targets and access to the greatest number and structural diversity of compounds. Traditional directed-synthesis chemical libraries, combinatorial chemistry libraries, and natural product extracts represent the most important pools of chemical entities available for HTS programs. Each of these compound pools brings different strengths to the drug discovery process. The directed synthesis libraries found in all major pharmaceutical companies have been accumulated through years of in-house medicinal chemistry programs. They contain synthetic intermediates on pathways to, and structural analogs of, biologically active target molecules. Advantages of lead compounds from these libraries are that the structures are already well characterized and a synthetic route to the compounds already exists. Combinatorial synthetic chemistry, which has only emerged in the last decade, has the potential to generate very large numbers of compounds and it is an extremely powerful tool for structure–activity optimization.[1,2,8] Initially, there was much enthusiasm for the idea that combinatorial chemistry could replace the need for accessing all other sources of chemical diversity, particularly natural product extracts. However, combinatorial libraries will always be designed with ease and efficiency of synthesis in mind, so they are likely to contain relatively simple structures. Living organisms, on the other hand, pay little attention to synthetic efficiency and, as a result, they contribute secondary metabolites with relatively complex structures, rich in stereochemistry, concatenated rings, and reactive functional groups, to the pool of chemical diversity.

The natural products chemical diversity pool, which is full of structural surprises unbiased by human imagination, has a strong track record of contributing to drug development. Terrestrial plant natural products (see Figures 1 and 2) that are currently in clinical use include the anti-cancer drugs paclitaxel (**3**) isolated from the Pacific Yew tree *Taxus* brevifolis and vincristine (**9**) isolated from the rosy periwinkle[9] *Catharanthus roseus*, the analgesic morphine (**10**) isolated from the poppy *Papaver somniferum* and the antimalarial drug artemisinin (**11**) isolated from the Chinese plant *Artemisica*

[7] J. R. Borach and J. Thorner, *Nature*, 1996, **384**, suppl., 14.
[8] J. C. Hogan, *Nature*, 1996, **384**, suppl., 17.
[9] H.-K. Wang, *IDrugs*, 1998, **1**, 92.

Figure 2 Terrestrial plant and soil microorganism secondary metabolites currently used as drugs

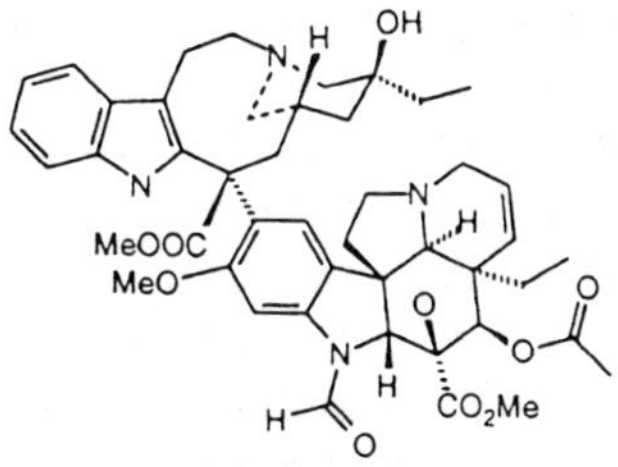

9 Vincristine

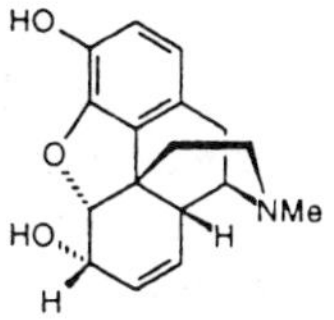

10 Morphine

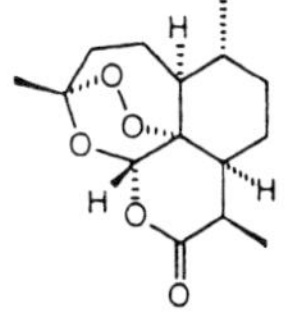

11 Artemisinin

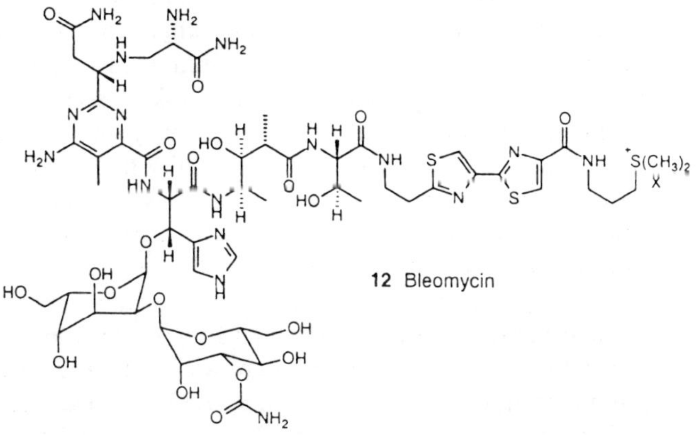

12 Bleomycin

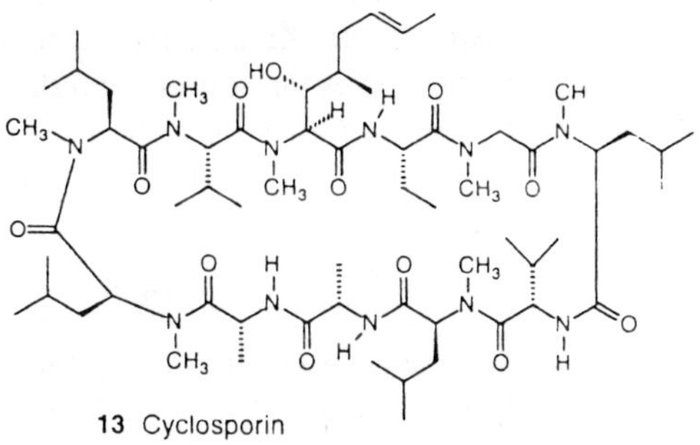

13 Cyclosporin

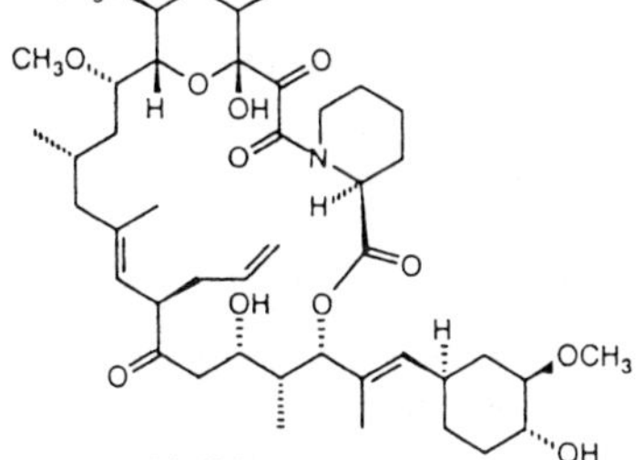

14 FK 506

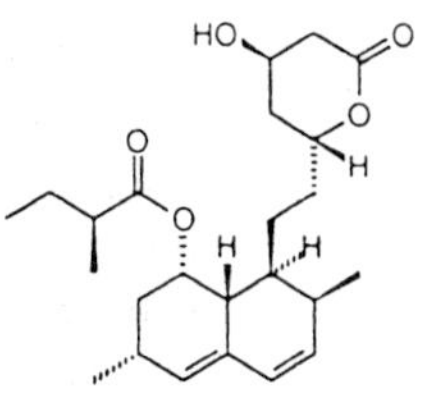

15 Lovastatin

annua.[10] Soil microbe natural products in clinical use include the anticancer drug bleomycin (**12**), the antibiotic vancomycin (**2**), the immunosuppresants cyclosporin (**13**) and FK 506 (**14**), and the cholesterol lowering agent lovastatin (**15**).[10] The near-term future of natural products in drug discovery also continues to look promising. A recent survey by National Cancer Institute scientists showed that natural products or synthetic analogs of natural products make up a significant proportion (*ca.* 50%) of anti-cancer and anti-infective 'lead' compounds currently in preclinical evaluation or clinical trials.[11] This promising pipeline of drug candidates almost guarantees that new, clinically useful agents based on natural product structures will continue to appear for the foreseeable future.

Despite the rich history of natural products contributing to drug development, the emergence of HTS as the predominant drug development paradigm has made most large pharmaceutical firms take a serious look at whether natural products have any long-term role to play in this new approach.[2,6] Their evaluations have focused on the possibility that the large libraries of chemical entities rapidly prepared by combinatorial chemistry have made natural product screening obsolete. In order to address this issue in an unbiased and systematic way, scientists at Bayer Pharma carried out a statistical analysis of the structural complementarity of natural products and synthetic compounds. Their results led to the conclusion that 'the potential for new natural products is not exhausted and natural products still represent an important source for the lead-finding process' in drug discovery.[12] What appears to be emerging, as a result of analyses such as that carried out by the Bayer group and the accumulation of several years worth of experience with the use of combinatorial libraries in HTS programs, is the recognition that although combinatorial chemistry is indeed an extremely powerful tool for drug discovery, it cannot completely replace natural products as a source of chemical diversity.

2 Opportunities in the Oceans

Secondary metabolites (natural products) are so named because they are apparently not essential for the primary metabolic activities involved in the growth of the producing organism. It is often argued that they act as antifeedants or toxins designed to reduce predation on the producing organism or that they act to inhibit the growth of other species competing for the same pool of nutrients or space. While these arguments are usually well founded, the reality is that for most secondary metabolites there is no rigorous experimental proof for their putative roles. In contrast to the almost universal occurrence of primary metabolites such as amino acids, fatty acids, and nucleotides in all organisms from microbes to mammals, the occurrence of any particular secondary metabolite or natural product is normally restricted to only one or at most a few species. It is also the case that not all living organisms make secondary metabolites. A high proportion of higher plants and microorganisms do make secondary metabolites while most vertebrates do not. The restricted taxonomic

[10] Y.Z. Shu, *J. Nat. Prod.*, 1998, **61**, 1053.
[11] G.M. Cragg, D.J. Newman and K.M. Snader, *J. Nat. Prod.*, 1997, **60**, 52.
[12] T. Henkel, H. Muller, R.M. Brunne and F. Reichel, *Angew. Chem., Int. Ed. Engl.*, 1999, **38**, 643.

distribution of secondary metabolites means that for phyla known to be rich in secondary metabolism, there is a rough correlation between species diversity and chemical diversity. Consequently, drug screening programs that are bioprospecting for maximum structural diversity of natural products must try to access the maximum available biodiversity.

Greater than 70% of the Earth's surface is covered by oceans. The oceans are much bigger volumetrically than inhabitable land surfaces so they provide much more space in which organisms can live and diversify. The thermal buffering capacity of seawater means that ocean temperatures are very stable in comparison to those of freshwater and terrestrial environments. Furthermore, the ions that contribute to the total salinity of seawater are in nearly the same proportions throughout the oceans and their total concentrations in seawater are very similar to their concentrations in physiological fluids, minimizing problems of osmoregulation and ionic regulation. The very stable physical and chemical environment in the oceans, the wide range of habitats extending from the very deep abyssal plain to very shallow tidal pools, and the ready availability of nutrients have been major factors in the development of the high biodiversity found in the marine environment.

Marine biota are an extremely exciting potential source of novel natural products not only because of the large number of species present in the oceans, but also because many of these plants and animals are uniquely adapted to marine habitats and have no terrestrial counterparts. Marine algae, for example, live immersed in seawater, which requires a set of physiological adaptations for accessing nutrients, exchanging metabolic gases, and reproduction that are very different from those found in terrestrial vascular plants. In spite of their physiological differences, one thing marine algae and terrestrial plants have in common is an apparent ecological need for biologically active secondary metabolites to deter herbivorous predators. Even though marine and terrestrial plants are significantly different from each other, it is among the invertebrates that there is the most striking difference between marine and terrestrial life forms. There are approximately 100 000 species of marine invertebrates in the world's oceans and the true number is likely much higher because there are still many undiscovered or undescribed species.[13] A number of the invertebrate phyla, including the Coelenterata, Porifera, Bryozoa, and Echinodermata, are exclusively aquatic and largely saline in habitat. Most of the animals in these phyla are sessile, slow growing, brightly colored, and lack the physical protection of shells or spines. Simple ecological arguments predict that organisms with exactly these attributes might make extensive use of biologically active secondary metabolites to defend themselves from predation by fishes, crabs, *etc.*

The field of marine natural products chemistry, which encompasses the study of the chemical structures and biological activities of secondary metabolites produced by marine plants, animals, and microorganisms, began in earnest in the early 1960s.[14,15] This is in stark contrast to the study of terrestrial plant natural

[13] R.C. Brusca and G.J. Brusca, *Invertebrates*, Sinauer Associates, Sunderland, MA, 1990, p. 4.

[14] P.J. Scheuer (ed.), *Bioorganic Marine Chemistry*, Springer, Heiderberg, 1987–92, vols. 1–6.

[15] D.J. Faulkner and R.J. Andersen, in *The Sea*, vol. 5, *Marine Chemistry*, ed. E.D. Goldberg, Wiley-Interscience, New York, 1974, ch. 19, p. 679.

products chemistry, which started almost a century and a half earlier with the isolation of the well-known alkaloids morphine, strychnine, and quinine. The relative inaccessibility of marine macroorganisms and difficulties associated with culturing marine microorganisms, particularly phytoplankton, probably contributed to the late bloom of marine natural products chemistry research. Cousteau's invention of SCUBA in 1943 dramatically increased the accessibility of sessile marine invertebrates and algae. With this tool in hand, it became practical for natural products chemists to make their own field collections of marine organisms. Once sample collection was conveniently under the chemist's control, the way was open for active investigation of marine secondary metabolism. Today, after nearly 40 years of research, there have been roughly 10 000 novel marine natural product structures described in the chemical literature and the rate at which they are reported has increased to *ca.* 600 per year in 1997 from *ca.* 250 per year in the decade 1977 to 1987.[16]

Several analyses of known marine natural products according to the phylum of the source organism have revealed the richest sources of marine secondary metabolism (Figure 3).[17,18] It is clear from the data that marine algae and invertebrates, which together account for more than 90% of all reported compounds to date, have been the richest sources of novel natural products in the oceans. This is in keeping with the rudimentary ecological arguments presented above that these two groups of organisms might be expected to utilize chemical defences to deter predators. Among the invertebrates, sponges, coelenterates (soft corals and gorgonians), echinoderms, tunicates, and molluscs have yielded the most compounds.

Interestingly, there has been a significant shift in emphasis during the last decade. In the early days of marine natural products research, algae accounted for *ca.* 40% of new compounds, invertebrates *ca.* 60%, and microorganisms *ca.* 1%. In the most recent analysis, algae account for only *ca.* 10% of the total, invertebrates account for *ca.* 80%, and microorganisms account for *ca.* 10%. Among the invertebrates, the proportion of sponge (26% in 1977–85; 43% in 1997) and tunicate (2.6% in 1977–85: *ca.* 10% in 1997) compounds have significantly increased, while the proportion of coelenterate compounds (22% in 1977–85; 13% in 1997) has decreased. These shifts probably reflect the recent focus of the field on drug discovery, particularly the active search for anti-cancer drugs leads. A review of cytotoxic marine natural products showed that sponges (44% of total) and tunicates (13% of total) were the most prolific marine sources of compounds with this particular type of bioactivity, so it is not surprising that interest in natural products from these phyla has increased.[19] The other major shift has been the increase in reports of microbial metabolites (1% of total in

[16] D. J. Faulkner, *Nat. Prod. Rep.*, 1998, **15**, 113; and the previous reports in this series.

[17] C. Ireland, D. Roll, T. Molinski, T. McKee, M. Zabriske and J. Swersey, in *Biomedical Importance of Marine Organisms*, ed. D. G. Fautin, California Academy of Sciences, San Francisco, 1988, p. 41.

[18] C. Ireland, B. Copp, M. Foster, L. McDonald, D. Radisky and C. Swersey, in *Marine Biotechnology*, vol. 1, *Pharmaceutical and Bioactive Natural Products*, ed. D. Attaway and O. Zaborsky, Plenum Press, New York, 1993, ch. 1, p. 1.

[19] F. J. Schmitz, B. F. Bowden and S. I. Toth, in *Marine Biotechnology*, vol. 1, *Pharmaceutical and Bioactive Natural Products*, ed. D. Attaway and O. Zaborsky, Plenum Press, New York, 1993, ch. 7, p. 197.

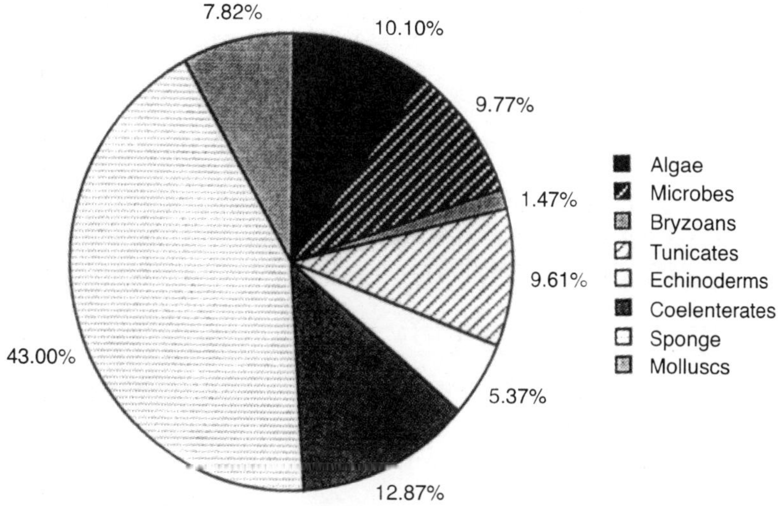

Figure 3 Distribution of marine natural products reported in 1997 according to the phylum of the source organism

1977–85: *ca.* 10% of total in 1997). This increase has been stimulated by the results of several pioneering studies showing that cultures of microorganisms uniquely adapted to the marine environment are a rich but essentially unexplored resource of novel bioactive secondary metabolites,[20–22] and parallel demonstrations that many of the exciting bioactive compounds isolated from marine invertebrate extracts are actually metabolites of associated microorganisms.[23]

The underlying assumption driving marine natural products chemistry research is that secondary metabolites produced by marine plants, animals, and microorganisms will be substantially different from those found in traditional terrestrial sources simply because marine life forms are very different from terrestrial life forms and the habitats which they occupy present very different physiological and ecological challenges. The expectation is that marine organisms will utilize completely unique biosynthetic pathways or exploit unique variations on well established pathways. The marine natural products chemistry research conducted to date has provided many examples that support these expectations.

Red algae were some of the first organisms to attract serious attention from marine natural product chemists.[24] The typical metabolites of red algae are mono-, sesqui-, and di-terpenoids and polyketides. What makes these compounds unique to marine plants is the high degree of substitution with chlorine and bromine atoms and the large number of unprecedented terpenoid carbon skeletons. Red algae in the genera *Laurencia* and *Plocamium* have been the richest source of halogentated algal terpenoids and polyketides. Some typical *Laurencia* terpenoids (see Figure 4) include pacifenol (**16**), perforatane (**17**), oppositol (**18**), and neoirieone (**19**), and some typical *Laurencia* polyketides include laurepinnacin

[20] W. Fenical, *Chem. Rev.*, 1993, **93**, 1673.

[21] B.S. Davidson, *Curr. Opin. Biotechnol.*, 1995, **6**, 284.

[22] V.S. Bernan, M. Greenstein and W.M. Maiese, *Adv. Appl. Microbiol.*, 1997, **43**, 57.

[23] C.A. Bewley and D.J. Faulkner, *Angew. Chem., Int. Ed. Engl.*, 1998, **37**, 2162.

[24] K.L. Erickson, in *Marine Natural Products, Chemical and Biological Perspectives*, ed. P.J. Scheuer, Academic Press, New York, 1983, vol. 5, ch. 4, p. 132.

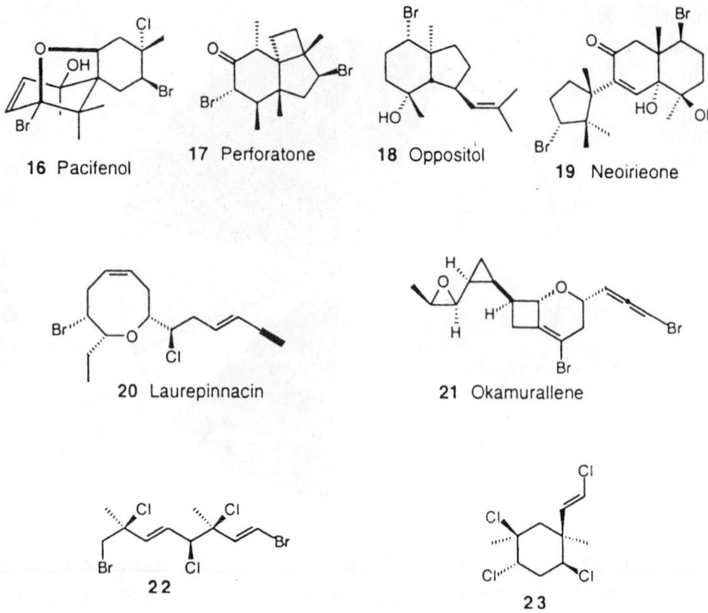

Figure 4 Halogenated terpenoid and polyketide metabolites isolated from red algae in the genera *Laurencia* and *Plocamium*

(**20**) and okamurallene (**21**). The dibromotrichloro acyclic compound **22** and the tetrachloro monocyclic compound **23** are typical *Plocamium* monoterpenoids.

It has been postulated that the biosynthesis of halogenated red algal metabolites involves a variation on the well-known terpenoid and polyketide pathways. The introduction of halogen atoms into the molecular skeletons is thought to arise from initial reaction of an electrophilic halonium ion (formally Br^+ or Cl^+) species with a nucleophilic alkene or alkyne bond to generate intermediate carbocations that can lose a proton, undergo ring cyclization reactions, or be trapped by bromide, chloride, or oxygen nucleophiles as shown in Figure 5 for the sesquiterpenoid nidificine (**24**). In terrestrial plant biosynthesis, the analogous cyclizations of acyclic terpenoid precursors are initiated by the electrophilic species H^+ rather than by halonium ions. The incorporation of halogen atoms into red algal metabolites is a typically marine biosynthetic variation on the basic terpenoid and polyketide pathways that takes advantage of the relatively high concentrations of chloride and bromide ions in seawater.

Dinoflagellates, a group of planktonic unicellular marine plants, make a remarkable group of extremely toxic polyketides (see Figure 6) that have no terrestrial counterparts.[25] Maitotoxin (**25**), produced by the dinoflagellate *Gambierdiscus toxicus*, is one of the toxins (LD = 50 ng/kg, mouse, ip) responsible for ciguatera seafood poisoning.[26] It is the largest (molecular weight = 3422 Daltons) non-biopolymer natural product known and it features a linear skeleton containing 142 carbon atoms that has been elaborated into multiple units of fused cyclic ethers. Okadaic acid and the related compound DTX-4 (**26**), biosynthetic products of dinoflagellates in the genera *Prorcentrum* and *Dinophysis*, are

[25] T. Yasumoto and M. Murata, *Chem. Rev.*, 1993, **93**, 1897.

[26] M. Sasaki, N. Matsumori, T. Maruyama, T. Nonomura, M. Murata, K. Tachibana and T. Yasumoto, *Angew. Chem., Int. Ed. Engl.*, 1996, **35**, 1672.

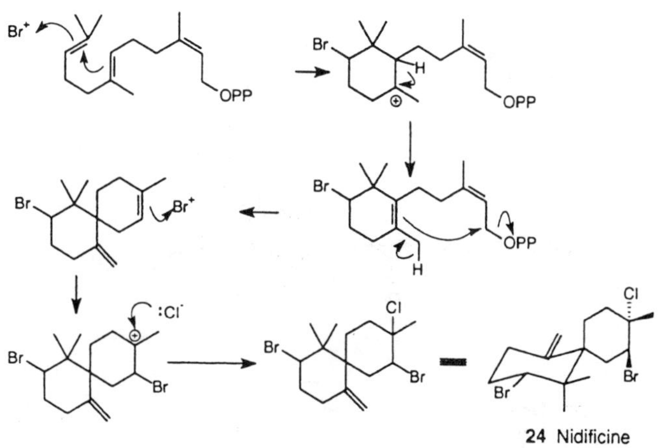

24 Nidificine

Figure 5 Proposed mechanism for biosynthetic introduction of bromine and chlorine atoms into the *Laurencia* sesquiterpenoid nidificine (**24**)

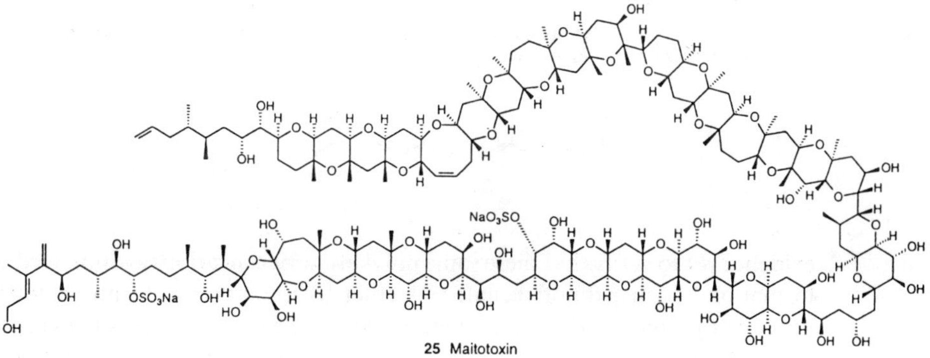

25 Maitotoxin

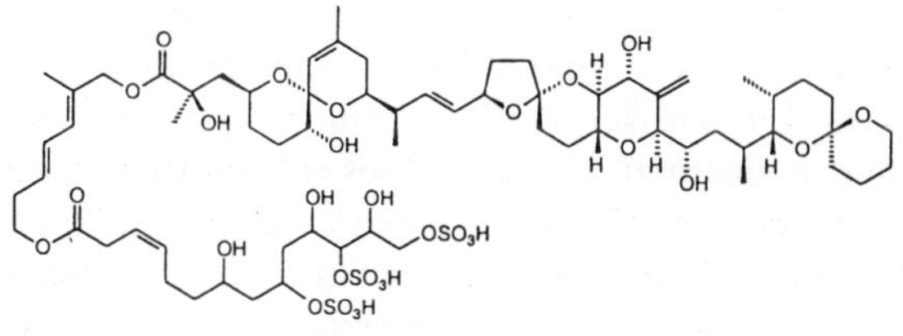

26 DTX-4

Figure 6 Toxic polyketide metabolites of marine dinoflagellates

members of the family of compounds that are responsible for causing diarrhetic shellfish poisoning.

Biosynthetic investigations of DTX-4 (**26**) have revealed that the molecule is formed via a polyketide pathway, but one that involves a number of unusual features.[27] These include the use of glycolate as a starter unit, the formation of pendant methyl groups and exomethylene groups from C-2 of acetate, cleavage of the polyketide chain via a Baeyer–Villiger-like oxidation, and deletion of individual carbons from the polyketide chain via a Favorskii-type rearrangement followed by decarboxylation as shown in Figure 7. While there is no experimental

[27] J. L. C. Wright, T. Hu, J. L. McLachlan, J. Needham and J. A. Walter, *J. Am. Chem. Soc.*, 1996, **118**, 8757.

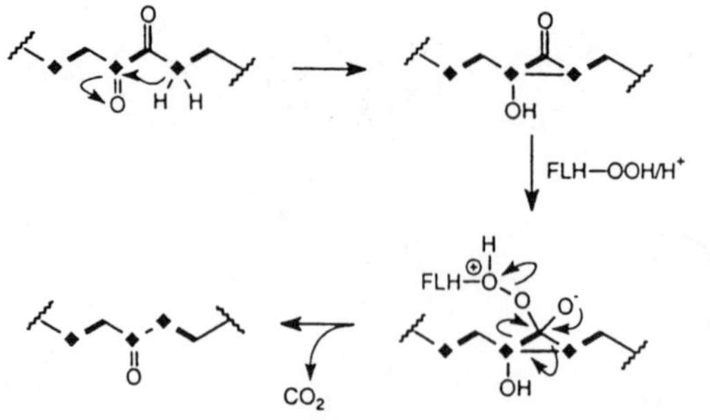

Figure 7 Proposed biological Favorskii-like reaction involved in DTX-4 biosynthesis. ◆— denotes incorporation of intact units of $^{13}CH_3$–$^{13}CO_2H$ in a biosynthetic feeding experiment

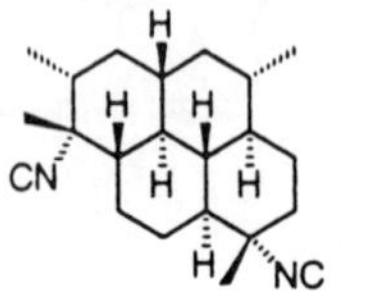

27 Diisocyanoadociane

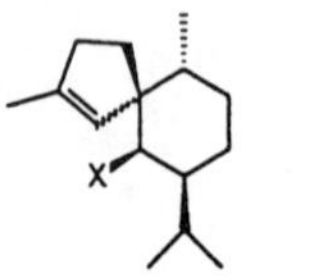

28 Axisonitrile-3 X = -NC
29 Axisothiocyanate-3 X = -NCS

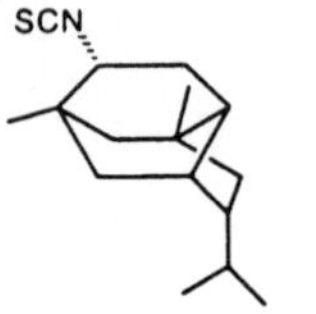

31 2-Thiocyanoneopupukeanane

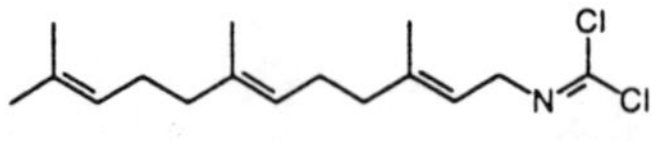

30

Figure 8 Sponge terpenoids containing the rare isocyanide, isothiocyanide, and dichloroimine functional groups

evidence yet to verify that these same polyketide modifications are involved in the formation of other dinoflagellate polyether toxins, structural features found in other compounds of this class suggest that these rare and interesting variations on the polyketide pathway may be a common feature of dinoflagellate biosynthesis.

Many marine invertebrate metabolites also have biosynthetic origins without known terrestrial counterparts. Two examples from marine sponges will serve as illustrations. Terpene isocyanides, isothiocyanides and dichloroimines are some of the most distinctive metabolites of marine sponges (see Figure 8). Diisocyanoadociane (**27**) from *Amphimedon terpenensis*, axisonitrile-3 (**28**) and axisothiocyanate-3 (**29**) from *Acanthella cavernosa*, the dichloroimine **30** from *Stylotella aurantium* and 2-thiocyanoneopupukeanane (**31**) from an *Axinyssa* sp. are representative examples. Biosynthetic feeding studies have shown that the isocyanide and isothiocyanide functionalities in these compounds arise from direct incorporation of inorganic cyanide and thiocyanate, presumably by reaction with terpene carbocation biosynthetic intermediates.[28] This is in direct contrast to the biosynthesis of the small number of isocyanide containing metabolites known from terrestrial microorganisms, where the isocyanide functionality arises from modification of an amino acid nitrogen.

Sponges in the order *Haplosclerida* have been the source of a family of more than 100 3-alkylpiperidine alkaloids (see Figure 9) that all appear to have been formed via a common biosynthetic pathway which is unique to marine

[28] J. S. Simpson and M. J. Garson, *Tetrahedron Lett.*, 1999, **40**, 3909.

Figure 9 3-Alkylpiperidine alkaloids isolated from sponges in the order *Haplosclerida*

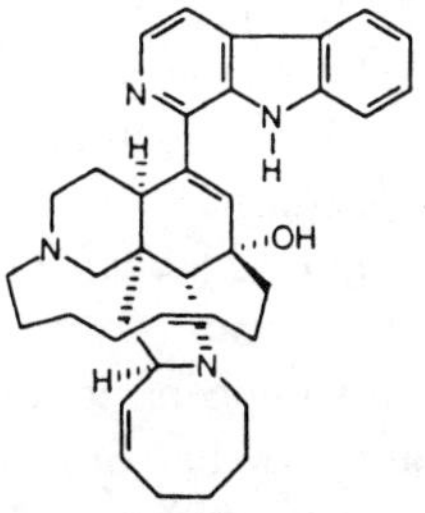

32 Manzamine A

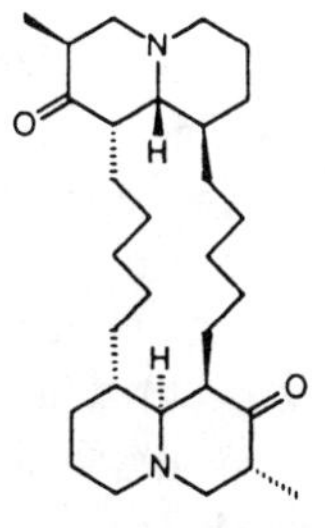

33 Petrosin A

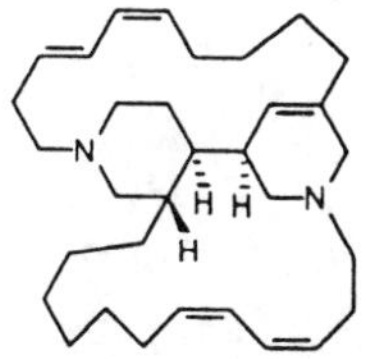

34 Halicyclamine A

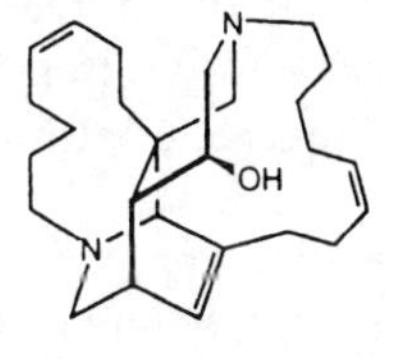

35 Ingenamine

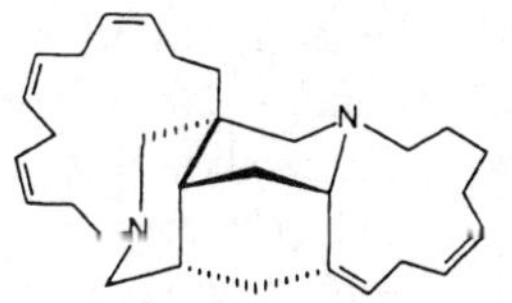

36 Madangamine A

Figure 10 Proposed biogenesis of 3-alkylpiperidine alkaloids

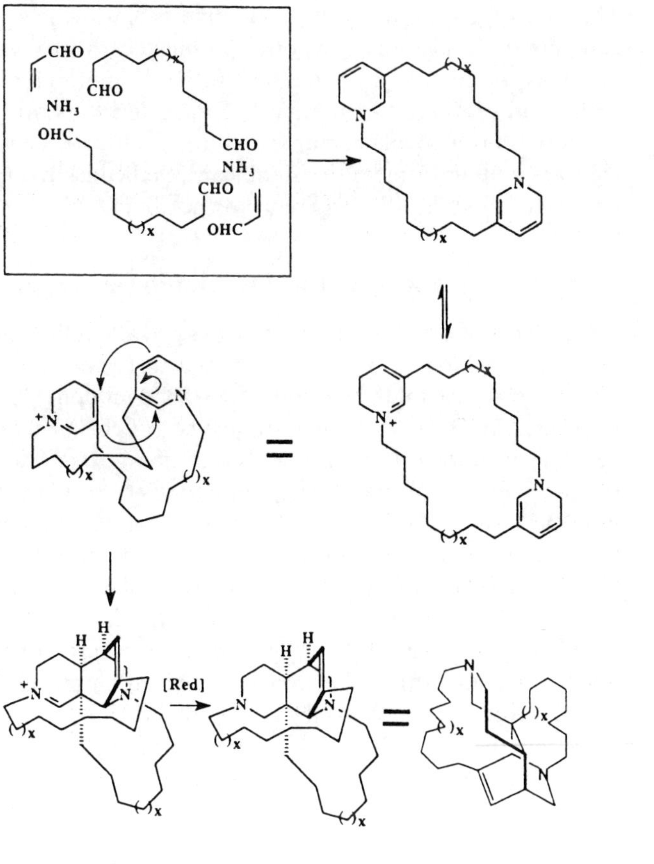

sponges.[29] Manzamine A (**32**) isolated from a *Haliclona* sp., petrosin A (**33**) isolated from *Petrosia seriata*, halicyclamine A (**34**) also isolated from a *Haliclona* sp., along with ingenamine (**35**) and madangamine A (**36**) isolated from *Xestospongia ingens* are representative examples.

It has been proposed that the basic building blocks involved in the biogenesis of all the 3-alkylpiperidine alkaloids are ammonia, a three-carbon unit such as propenal, and a long-chain dialdehyde as shown in Figure 10. These units are first assembled into a 3-alkylpiperidine monomer that can either polymerize to give the oligomeric halitoxins or dimerize to give a bis(3-alkylpiperidine) macrocycle. Bis(3-alkylpiperidine) macrocycles can undergo various transformations, including intramolecular [4 + 2] cycloaddition reactions, to generate the polycyclic skeletons found in many of these complex alkaloids. There is no terrestrial counterpart to the proposed biogenetic pathway to these complex 3-alkylpiperidine sponge alkaloids.

The selected examples described above serve to illustrate that marine natural products are indeed significantly different from the natural products traditionally isolated from terrestrial plants and microorganisms. Numerous excellent reviews of the marine natural products literature are full of many additional examples. This marine pool of novel chemical diversity, which contains many remarkably complex molecules that are often imbued with exquisitely potent biological activities, represents an extremely valuable molecular resource for the 'lead finding' process in contemporary HTS drug discovery programs. The opportunity to exploit this valuable resource and the resulting rewards are waiting for those who are willing to meet the significant challenges required to bring a marine natural product 'lead' compound to the clinic.

3 Challenges Involved in Developing a 'Drug from the Sea'

The process of developing any new drug is a lengthy and very costly business. Typically it takes 12–15 years to bring a drug to market from the time of its initial discovery and the total bill is more than $350 million (US). The failure rate is also enormous. In the initial stages, thousands of chemical entities are evaluated to find one or more 'lead' compounds that merit extra evaluation and only a very small percentage of these 'lead' compounds ever make it into the clinic. These timeline, cost, and failure rate barriers to drug development exist regardless of whether the 'lead' compound comes from a synthetic or natural product chemical diversity pool. However, there are a number of unique challenges to developing a drug based on a natural product from a marine plant or animal source.

The first step in discovering a marine natural product 'lead' compound involves accumulating collections of the source organisms. Ideally the collections would include as many species as possible collected from a large number of sites reflecting the wide diversity of habitats found in the world's oceans. The development of SCUBA and both manned and remotely operated (ROV) submersibles has provided the technology required to effectively collect algae and invertebrate specimens from most marine habitats. Consequently, the major issue

[29] R. J. Andersen, R. W. M. Van Soest and F. Kong, in *Alkaloids: Chemical & Biological Perspectives*, ed. S. W. Pelletier, Pergamon, New York, 1996, vol. 10, ch. 3, p. 301.

in collecting marine specimens now involves securing permission to collect from the countries whose territorial waters encompass the desired collecting sites.

There are two issues that must be addressed when securing collecting permits. The first relates to the possible environmental impact of collecting marine specimens. Many jurisdictions require collecting permits that are much like fishing licenses in an attempt to control the excessive removal of marine specimens by SCUBA, dredging, or submersibles. Sometimes the regulations cover entire coastlines, other times they only cover protected areas like marine parks and reserves or high profile tourist destination dive sites. Fortunately, modern HTS assays require very small quantities of crude extracts, so initial collections of specimens for screening purposes can usually be limited to less than 500 g. Recent advances in isolation and structure elucidation technology have also dramatically reduced the quantities of crude extract required to identify the chemical structures of bioactive hits. Therefore, the total biomass of a specimen currently required for HTS screening and compound identification is quite small and collection of this quantity should normally have no environmental impact. A carefully documented case along these lines is usually adequate to obtain a collecting permit where environmental issues are the concern.

The second, more complex issue involved in obtaining collecting permits has to do with intellectual property rights, licensing fees, and royalties. Two international documents deal with this issue. The first is the Convention of Biological Diversity that was signed by 140 nations at the 1992 Earth Summit in Rio de Janeiro (as of 1997, only 30 countries have ratified the document). In the convention, ethical concerns regarding the intrinsic value of biodiversity were addressed, but the main focus was to establish sovereign rights over biological resources, developing mechanisms for their sustainable use, and ensuring equitable distribution of the resulting benefits.[30] The convention acknowledged that, in general, international assistance would have to be provided to developing countries to pursue biodiversity protection efforts, and acknowledged the rights of countries to obtain compensation for access to their genetic resources, including transfer of relevant technologies and equitable sharing of research results and other benefits. However, many of the provisions of the convention are quite vague and contradictory, particularly with regard to intellectual property rights and the compensation arrangements that should be applied to transactions involving genetic resources. Essentially, countries are left to decide for themselves what constitutes compliance.[30]

TRIPS, the Trade Related Aspects of Intellectual Property Rights agreement of the World Trade Organization, which came into effect in 1995, is the second international document that deals with intellectual property rights arising from biodiversity. This document is in sharp conflict with the basic premise endorsing the concept of nations holding property rights to their indigenous species as outlined in the Rio Accord. The TRIPS assumptions are that excessive government control of resources is to be avoided as an obstacle to economic growth. Such obstacles include the ability of individuals to negotiate market

[30] A. Artuso, *Drugs of Natural Origin: Economic and Policy Aspects of Discovery, Development, and Marketing*, The Pharmaceutical Products Press, New York, 1997, p. 142.

prices of biological resources that the Biodiversity Convention allows.[31] The Organization of African Unity's task force on access to genetic resources has argued that TRIPs should comply with the Biodiverity Convention, and not the other way round, and that member countries may refuse to recognize any patent from natural products found in Africa unless it acknowledges the 'ownership' and contribution of the relevant community to the new product.[31]

The National Institute of Health (NIH) in the USA has taken a leadership role in this area by making sure that collections of marine specimens made in foreign waters using NIH funds are conducted only under agreements that recognize the rights of the country of origin of the biological materials. The responsibilities of the collectors and the benefits to the country of origin will usually be formalized in a Memorandum of Understanding (MOU). MOUs spell out how the collections and ensuing research will be conducted and how proceeds from intellectual property rights will be shared. There is an emphasis on involving local scientists in the research, promoting technology transfer, and encouraging educational exchanges as well as negotiating equitable formulas for sharing licensing fees and royalty payments.

Once extracts have been obtained, they have to be screened for bioactivity and the chemical structures of the active constituents have to be elucidated. Contemporary HTS programs have been designed to speed up the process of identifying 'lead' structures in an attempt to shorten the timeline for drug development. Campaigns will often be mounted to screen full libraries of up to several hundred thousand compounds against a particular molecular target with the intent of identifying a complete suite of 'lead' structures in a period of no more than a few weeks or months. Combinatorial and directed synthetic libraries are ideally suited to this purpose, because they provide immediate 'lead' structures once hits have been identified by HTS. In contrast, a hit in a natural product crude extract provides no immediate structural information and, furthermore, the time required for bioassay-guided fractionation and structure elucidation, which can be days, months, or years, is highly variable from extract to extract. The delay in providing an active structure and the unpredictability of the process are hurdles that natural product extracts must overcome to be compatible with HTS. Ideally the time for assay-guided fractionation and chemical structure elucidation must be shortened to a few weeks or probably, at most, three months. Technological advances in NMR, MS, and HPLC, and the emergence of powerful hyphenated analytical techniques such as HPLC-MS, HPLC-NMR, and HPLC-CD,[32] have partially solved this problem by greatly accelerating the rate of routine structure elucidation. Other approaches to solving this problem involve preliminary purification of crude extracts before they are submitted to HTS and the creation of libraries of purified and identified natural products that can then compete directly with synthetic libraries.[33]

After a bioactive marine natural product 'lead' structure has been identified,

[31] C. MacLlwain, *Nature*, 1998, **392**, 535: *ibid.*, 1998, **392**, 525; E. Masood, *ibid.*, 1998, **392**, 537.

[32] G. Bringmann, K. Messer, M. Wohlfarth, J. Kraus, K. Dumbuya, M. Ruckert, *Anal. Chem.*, 1999, **71**, 2678.

[33] C. Koch, T. Neumann, R. Thiericke and S. Grabley, in *Drug Discovery from Nature*, ed. S. Grabley and R. Thiericke, Springer, Berlin, 1999, ch. 3, p. 51.

either by HTS or other traditional screening approaches, the issues of short-term small-scale and long-term manufacturing-scale supply become the determining factors in its further progress towards development. In the short term, 10s to 100s of milligrams of the compound are required to test for activity in secondary assays that probe such issues as activity in whole cells, *in vivo* efficacy in animal models, maximum tolerated doses and minimum effective doses in animals, and pharmacokinetics. Demonstration of *in vivo* activity is an absolute requirement before development can proceed.

Obtaining 10s or 100s of milligrams of many marine natural products for secondary assays represents a considerable challenge. This usually involves recollection of significant quantities of the source organism and the environmental impact of these larger collections must be carefully evaluated since many of the source organisms are rare species. The shift of emphasis towards marine microorganisms is partly a result of the supply issues since laboratory fermentation of a marine microorganism represents a sustainable supply of an interesting metabolite. It also reflects the emerging recognition that many interesting compounds isolated from marine invertebrate extracts are actually biosynthetic products of associated microorganisms.[23] The ability to culture the producing microorganism would provide a suitable supply of the desired compound both for secondary assays and ultimate commercial production.

Moving a promising marine natural product from the demonstration of *in vivo* activity in secondary assays to a genuine drug development project is a major step because of the required commitment of both financial and human resources. In order for a large pharmaceutical company to proceed, they must first determine that there is potential market niche for the compound that will return their development investment and, second, they must be assured of a reliable and economical source of the compound. Organic synthesis represents the ideal industrial source of a drug. Many marine natural products with promising *in vivo* activity have very complex chemical structures, making them attractive targets for academic synthetic programs. While academic syntheses prove that a particular compound can be made in the laboratory and often provide valuable structure–activity data, few of them are short enough or use reagents that are cheap or safe enough to become economical large-scale industrial syntheses.

For marine natural products that are too complex to be produced commercially by total synthesis, there are still several supply options. The first option is to use the natural product as a 'lead' structure for a synthetic medicinal chemistry program that identifies simpler synthetically accessible analogs that contain the same pharmacophore. Biological production of the natural product via fermentation or aquaculture represents a second potential solution to the supply problem. Many drugs are produced by industrial-scale fermentations of soil microorganisms, so conceptually this should be straightforward for a metabolite of a marine microorganism. The only difference is the need to solve the technical problems associated with preparation of large volumes of seawater-based fermentation medium and the incompatibility of the high concentration of salts in the medium with currently available stainless-steel fermentation equipment. Industrial-scale production of complex marine invertebrate metabolites by aquaculture of the producing organism under controlled conditions is an exciting prospect. While this technology is still in its infancy, the feasibility of the approach has recently been demonstrated by the production of anti-cancer compounds in the bryostatin family using controlled cultures of the bryozoan *Bugula neritina*.[34]

4 Some Success Stories

The number of marine natural product derived drugs currently in clinical use is still very small; however, there are many promising marine natural products or synthetic analogs in clinical or pre-clinical trials, so the expectation that new drugs based on marine natural products will appear in the clinic in the next few years is genuine. Two products (see Figure 11) based on marine natural products that have already reached the market are the anti-cancer drug ara C (**37**) and the anti-inflammatory pseudopterosin diterpenoids [*i.e.* pseudopterosin A (**38**)]. Ara C (**37**) (trade name Cytarabine) is a synthetic material based on the 'lead structures' provided by isolation of spongothymidine (**39**) and spongouridine (**40**) from the sponge *Cryptotethya crypta* in the 1950s.[35] It is currently the most active anti-metabolite in clinical use for inducing remissions in acute non-lymphocytic leukemia in adults and pediatric patients.[36] In combination with other anti-cancer agents, such as anthracyclines, ara C treatment leads to complete remission in 60–70% of treated patients. It can be viewed as a first generation 'Pharmaceutical from the Sea' that emerged from the classical model of using a bioactive natural product as a 'lead structure' for the development of a synthetic drug.

The pseudopterosins, isolated from the Caribbean soft coral *Pseudopterogorgia elisabethae*, are potent inhibitors of inflammation and its associated pain.[37] They have potential applications in a number of conditions including arthritis, psoriasis, and inflammatory bowel disease. The skin care company Estée Lauder is currently using a 'pseudopterosin-containing' natural extract of *P. elisabethae* in a product line called 'Resilience'. The natural extract produces reduction of skin inflammation from sunburn or irritation, and it also reduces degeneration of the skin. In this instance, the ability to obtain the active pseudopterosins as part of a 'natural marine extract' as opposed to making them 'synthetically' appears to be an important part of the marketing strategy.

Manoalide (**41**) was originally isolated from the Indo-Pacific sponge *Laffariella variabilis* and its chemical structure was reported without any mention of biological activity.[38] Subsequently, it was discovered that manoalide had exciting anti-inflammatory properties.[39] The anti-inflammatory activity of manoalide apparently results from its ability to inhibit the enzyme phospholipase A_2 (PLA_2). This ubiquitous enzyme catalyzes the hydrolysis of the ester in the *sn*-2 position of phospholipids, resulting in the release of arachadonic acid, the biosynthestic precursor of prostaglandins and leukotrienes in the arachadonic

[34] M. Jaspars, *Chem. Ind. (London)*, 1999, January, p. 51.
[35] W. Bergmann and D.C. Burke, *J. Org. Chem.*, 1956, **21**, 226.
[36] V. I. Avramis, in *A Clinician's Guide to Chemotherapy, Pharmacokinetics and Pharmacodynamics*, ed. L. B. Growchow and M. M. Ames, Williams and Wilkins, Baltimore, 1998, ch. 9, p. 209.
[37] A. M. S. Mayer, P. B. Jacobson, W. Fenical, R. S. Jacobs and K. B. Glaser, *Life Sci.*, 1998, **62**, 401.
[38] E. D. de Silva and P. J. Scheuer, *Tetrahedron Lett.*, 1980, **21**, 1611.
[39] B.C. Potts, D.J. Faulkner and R.S. Jacobs, *J. Nat. Prod.*, 1992, **55**, 1701.

Figure 11 Anti-inflammatory or anti-cancer marine natural products and synthetic analogs that are in commercial use or in clinical and preclinical evaluation

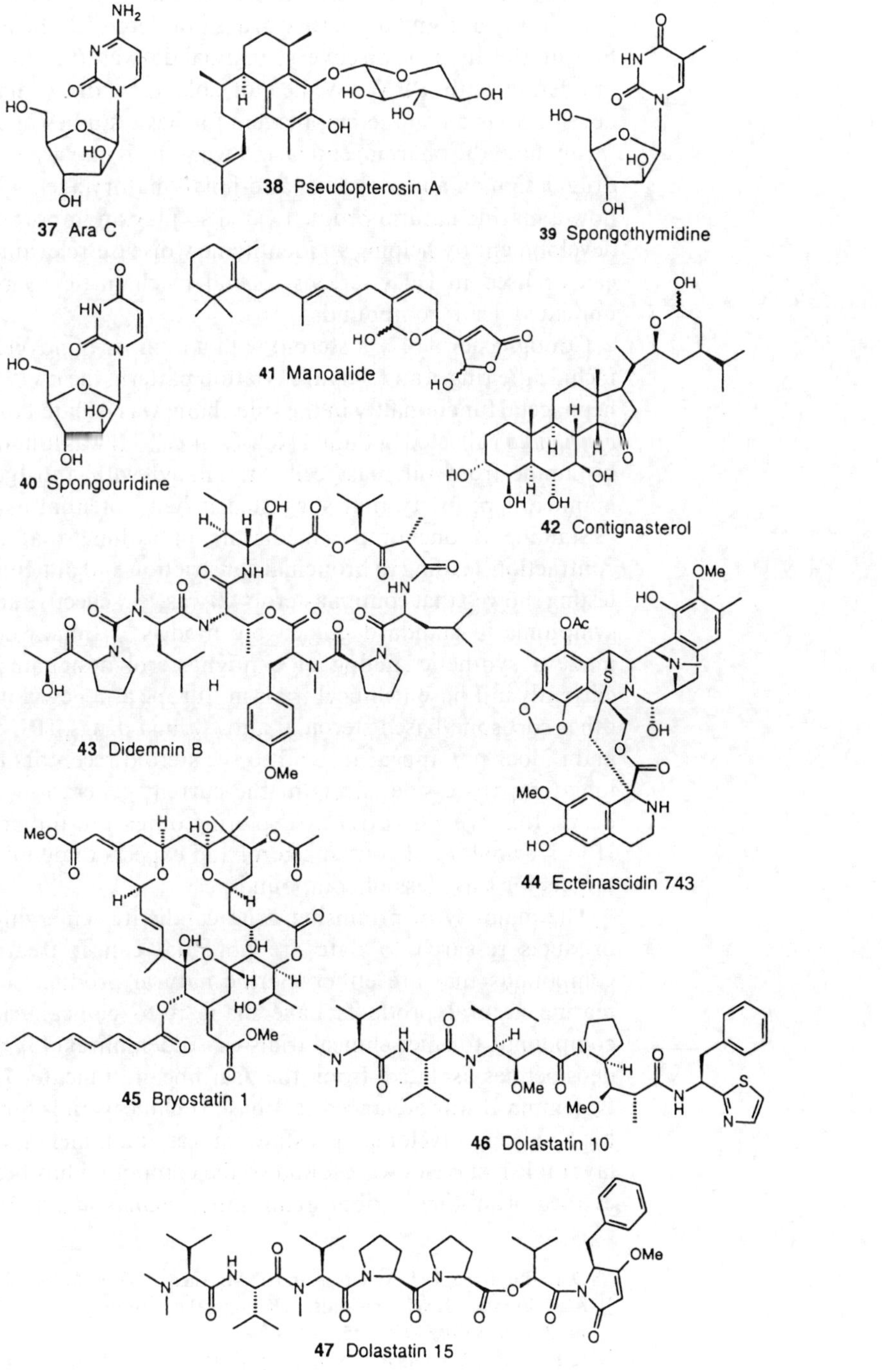

acid cascade. The prostaglandins are well-known mediators of inflammation and pain. Many anti-inflammatory drugs interfere with the archadonic acid cascade but, at the time of discovery, manoalide was the only one that worked by interfering with PLA_2. While not sold as a drug, manoalide (**41**) has been developed as a commercial reagent for basic studies of PLA_2. It is widely used throughout the pharmaceutical industry and it may well result, through synthetic drug design, in a new class of anti-inflammatory agents. It is a good example of how a marine natural product can also play an important indirect role in drug development by helping to identify new disease-relevant molecular targets that can be used in HTS screens to find mechanistically related but structurally unrelated 'lead' compounds.

Contignasterol (**42**), a steroid with a number of novel structural features that include the ring A and B hydroxylation pattern, the *cis* C/D ring junction, and the hemiacetal functionality in the side chain, was isolated from the sponge *Petrosia contignata* collected in Papua New Guinea.[40] It was found to inhibit the release of histamine from rat mast cells stimulated with anti-Ige in a dose-dependent manner, a property that suggested it had potential as an anti-asthma agent. Histamine is one of the mediators of asthma that causes smooth muscle contraction, leading to bronchial constriction and difficulty in breathing. Further testing showed that contignasterol (**42**) was very effective at preventing asthma-like symptoms in standard guinea pig models.[41] Inflazyme Pharmaceuticals has made a synthetic analog of contignasterol designated IPL-576-092, that is currently in Phase I clinical trials in Europe as a new anti-asthma agent. Unlike other cortisone-based steroidal anti-asthma drugs, IPL-576-092 is orally active and it does not appear to bind to the steroid receptors that are responsible for long-term toxic side effects in the current generation of inhaled anti-asthma drugs. Inflazyme is currently exploring other anti-inflammatory indications for IPL-576 analogs of contignasterol and expects to define additional therapeutic targets for this class of compounds.

The majority of promising drug candidates emerging from marine natural products research to date are potential cancer treatments. Six anti-cancer compounds that are either marine natural products or synthetic analogs of marine natural products have made it to clinical trials. The first of these compounds to enter clinical trials was didemnin B (**43**), one of a family of cyclic depsipetides isolated from the Caribbean tunicate *Trididemnum solidum.*[42] Didemnin B was advanced to Phase II clinical trials for treatment of small cell lung cancer, myeloma, prostate cancer, and melanoma. Unfortunately, no favorable responses were found so the compound has been withdrawn.[43] Crude extracts of another Caribbean tunicate, *Ecteinascidia turbinata*, showed extremely

[40] D. L. Burgoyne, R. J. Andersen and T. M. Allen, *J. Org. Chem.*, 1992, **57**, 525.

[41] A. M. Bramley, J. M. Langlands, A. K. Jones, D. L. Burgoyne, Y. Li, R. J. Andersen and H. Salari, *Br. J. Pharmacol.*, 1995, **115**, 1433.

[42] K. L. Rinehart, T. G. Holt, N. L. Fregeau, P. A. Keifer, G. R. Wilson, T. J. Perun, R. Sakai, A. G. Thompson, J. G. Stroh, L. S. Shield, D. S. Seigler, D. H. Li, D. G. Martin and G. Gade, *J. Nat. Prod.*, 1990, **53**, 771.

[43] For one example, see: S. K. Williamson, M. K. Wolf, M. A. Eisenberger, M. O'Rourke, W. Brannon and E. D. Crawford, *Invest. New Drugs*, 1995, **13**, 167.

promising tumor inhibitory properties. Two groups simultaneously succeeded in isolating and elucidating the structures of members of the ecteinascidin family of tetrahydroisoquinoline alkaloids that were responsible for the cytotoxicity of the crude extracts.[44,45] One member of this family of compounds, ecteinascidin 743 (**44**), is currently in clinical trials.[10] The material required for clinical trials was obtained from harvest of wild specimens.

Two additional marine natural product structural classes that have led to anti-cancer drug candidates in clinical trials have come from bryozoan and mollusc extracts. The bryostatins are a family of potently cytotoxic macrolides isolated from the bryozoan *Bugula neritina*.[46] Bryostatin 1 (**45**) is currently in late Phase I and early Phase II clinical trials in Europe for the treatment of melanoma. The supply of bryostatin 1 (**45**) required for preclinical testing came from harvest of wild specimens; however, the generally low yield and large variability from collection to collection make this a non-viable source for commercial production should a compound in this family make it to the clinic. An aquaculture approach to producing the bryostatins that is capable of generating a sufficient and sustainable supply for clinical use has been developed by CalBioMarine. The dolastatins are a family of cyclic and linear peptides isolated from the sea hare *Dolabella auricularia* collected in the Indian Ocean.[46] Isolation and structure elucidation of the potently cytotoxic dolastatins was hampered by their very low yields in the *D. auricularia* crude extracts. Several large collections of the animals that exceeded 1500 kg wet weight in total were processed in order to obtain the milligram quantities of individual compounds required for structure elucidation and preliminary biological testing. Subsequently, many of the dolastatins and a wide variety of structural analogs have been synthesized in order to confirm the structures of the natural product and provide material for further biological evaluation. Two linear peptides, dolastatin 10 (**46**) and dolastatin 15 (**47**), which have been found to be potent anti-mitotic agents that inhibit the polymerization of tubulin, have emerged as the most promising 'lead' compounds in the family. Dolastatin 10 (**46**) and a synthetic analog auristatin PE, along with dolastatin 15 (**47**) and a water soluble synthetic analog LU-103793, are all in phase I clinical trials.[10,47]

In addition to the compounds that have already moved to clinical trials, there are a number of very exciting marine natural products (see Figure 12) that are in pre-clinical evaluation as potential anti-cancer agents. Many of these turn out to be anti-mitotic agents that interfere with tubulin dynamics either by inhibiing polymerization or by stabilizing microtubules. Halichondrin B (**48**), a complex macrolide initially isolated from the Japanese sponge *Halichondria okadai*, is a potent inhibitor of tubulin polymerization that shows very promising *in vivo*

[44] K. L. Rinehart, T. G. Holt, N. L. Fregeau, J. G. Stroh, P. A. Keifer, F. Sun, L. H. Li and D. G. Martin, *J. Org. Chem.*, 1991, **56**, 1676.

[45] A. E. Wright, D. A. Forleo, G. P. Gundawardana, S. P. Gunasekera, F. E. Koehn, O. J. McConnell, *J. Org. Chem.*, 1990, **55**, 4508.

[46] G. R. Pettit, *J. Nat. Prod.*, 1996, **59**, 812–821.

[47] H. C. Pitot, E. A. McElroy, J. M. Reid, A. J. Windebank, J. A. Sloan, C. Erlichman, P. G. Bagniewski, D. L. Walker, J. Rubin, R. M. Goldberg, A. A. Adjeu and M. M. Ames, *Clin. Cancer Res.*, 1999, **5**, 525.

Figure 12 Some promising marine natural product 'lead' compounds

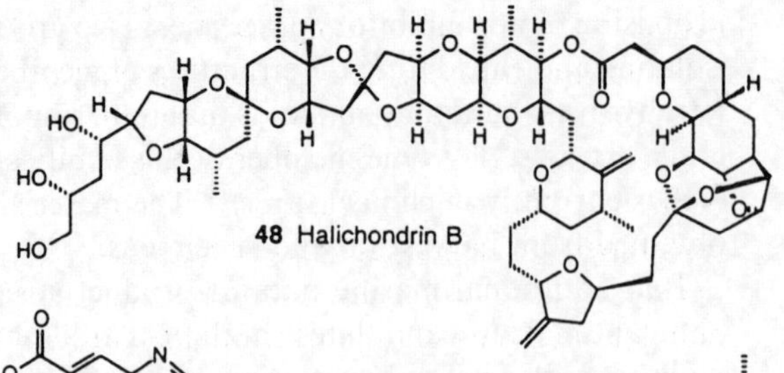
48 Halichondrin B

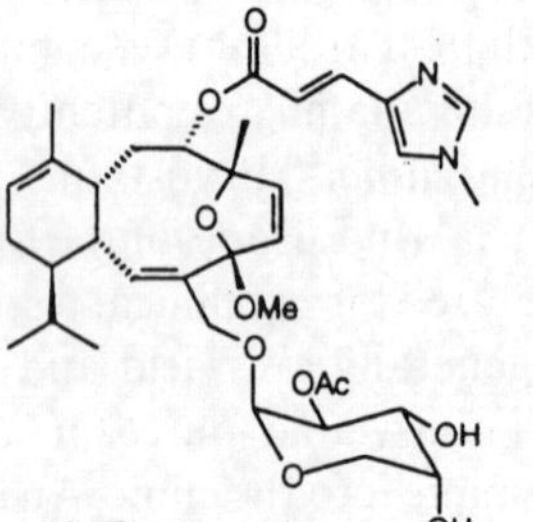
49 Eleutherobin

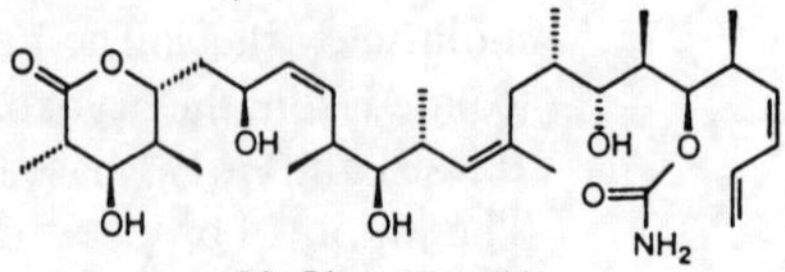
50 Discodermolide

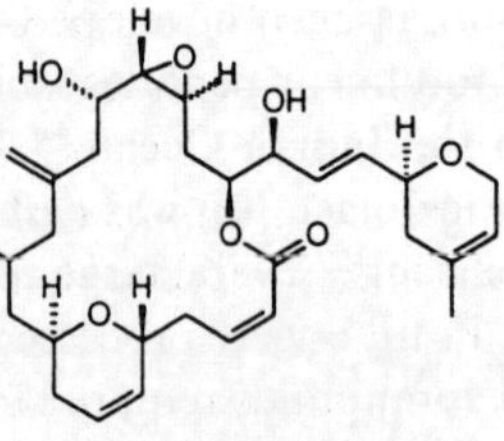
51 Laulimalide

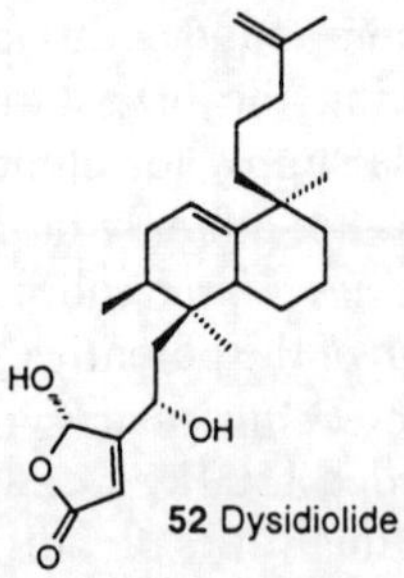
52 Dysidiolide

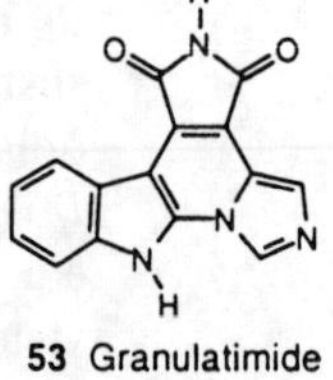
53 Granulatimide

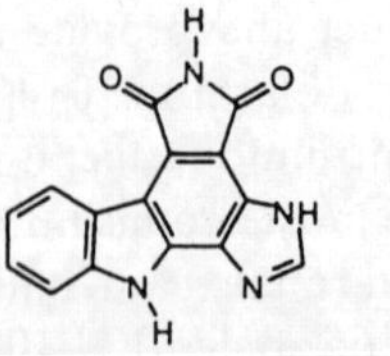
54 Isogranulatimide

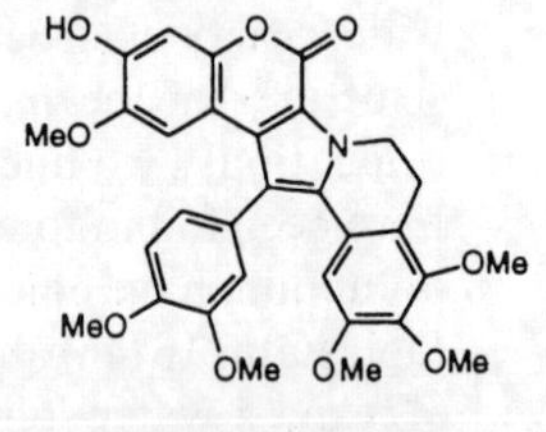
55 Lamellarin I

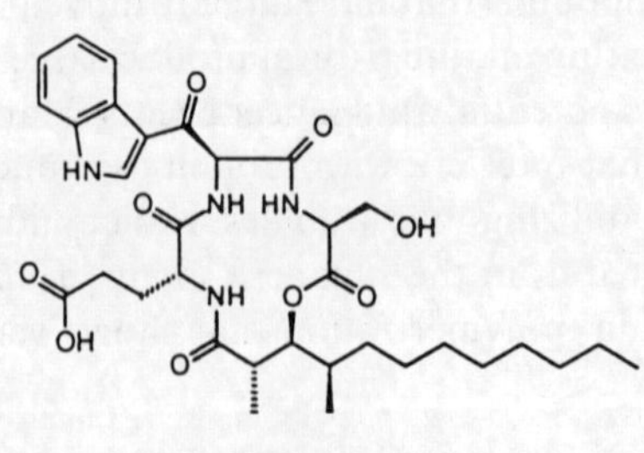
56 15G256γ

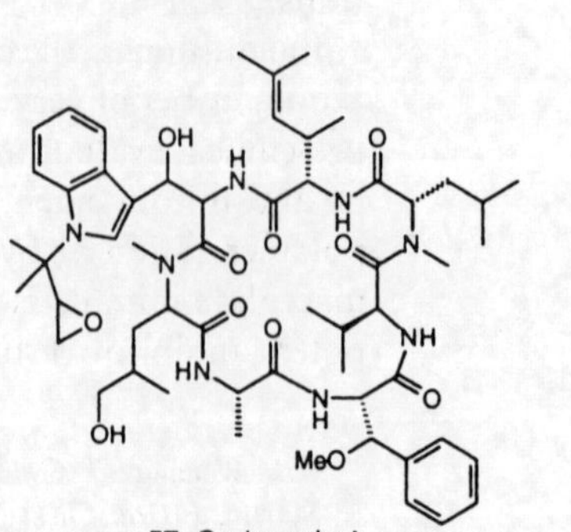
57 Cyclomarin A

activity against a number of human cancer cell lines in mouse xenografts.[48] The promising anti-tumor activity, but very limited supply, of halichondrin B has prompted the development of a method of producing the compound by *in situ* aquaculture of the New Zealand sponge *Lissodendoryx*. The recent introduction of the plant natural product paclitaxel into clinical use has prompted many researchers to search for other natural product structural classes that, like paclitaxel, also kill cells by stabilizing microtubules. Eleutherobin (**49**), a diterpenoid glycoside isolated from the Western Australian soft coral *Eleutherobia* sp.,[49] discodermolide (**50**), a polyketide isolated from the Caribbean sponge *Discodermia dissoluta*,[50] and laulimalide (**51**), isolated from the South Pacific sponge *Cacospongia mycofijiensis*,[51] are three novel marine natural products that mimic paclitaxel's ability to stabilize microtubules. All three compound classes are currently the focus of extensive chemical and biological research aimed at evaluating their clinical potential.

The anti-inflammatory and anticancer 'lead' compounds discussed above were all discovered by screening programs that used traditional whole cell or tissue-based assays and all come from extracts of marine invertebrates. Manoalide and the pseudopterosins were discovered using a mouse ear edema assay and all of the anti-cancer compounds were initially isolated because they showed potent *in vitro* cytotoxicity against either murine or human cancer cell lines. The recent shift in screening paradigms towards more focused molecular targets, using purified enzymes and receptors or engineered cell lines, is also taking place in marine natural products drug discovery research. This has resulted in reports of some exciting new activities for marine natural product chemotypes.

Dysidiolide (**52**) isolated from the Caribbean sponge *Dysidea etheria* is the first naturally derived inhibitor of cdc25A protein phosphatase, an enzyme involved in the activation of the G2/M transition of the cell cycle.[52] Dysidiolide shows micromolar *in vitro* inhibition of A-549 human lung carcinoma, presumably owing to its inhibition of cdc25A, making it an exciting 'lead compound' for development of a mechanism-of-action-based anti-cancer drug. Granulatimide (**53**) and isogranulatimide (**54**), metabolites of the Brazilian tunicate *Didemnum granulatum*, are the first G2 cell-cycle checkpoint inhibitors discovered by rational screening.[53] When used in combination with either a chemical DNA damaging agent or radiation, alkaloids **53** and **54** enhance the killing of cancer cells, missing the tumor suppressor gene (*i.e.* p53 – cells) relative to cells that

[48] O. Fodstad, K. Breistol, G. R. Pettit, R. H. Shoemaker and M. R. Boyd, *J. Exp. Ther. Oncol.*, 1996, **1**, 119.

[49] T. Lindel, P. R. Jensen, W. Fenical, B. H. Long, A. M. Casazza, J. Carboni, C. R. Fairchild, *J. Am. Chem. Soc.*, 1997, **119**, 8744.

[50] E. ter Haar, R. J. Kowalski, E. Hamel, C. M. Lin, R. E. Longley, S. P. Gunasekera, H. S. Rosenkranz and B. W. Day, *Biochemistry*, 1996, **35**, 243.

[51] S. L. Mooberry, G. Tien, A. H. Hernandez, A. Plubrukarn and B. S. Davidson, *Cancer Res.*, 1999, **59**, 653.

[52] S. P. Gunasekera, P. J. McCarthy, M. Kelly-Borges, E. Lobovsky and J. Clardy, *J. Am. Chem. Soc.*, 1996, **118**, 8759.

[53] M. Roberge, R. G. S. Berlinck, L. Xu, H. Anderson, L. Lim, D. Curman, C. M. Stringer, S. H. Friend, P. Davies, I. Vincent, S. J. Haggarty, M. T. Kelly, R. Britton, E. Piers and R. J. Andersen, *Cancer Res.*, 1998, **58**, 5701.

contain the tumor suppressor gene (*i.e.* p53+ cells). This combination therapy approach offers the hope of a greatly improved therapeutic index for the approximately one half of all human solid tumors that are known to be missing the tumor suppressor gene p53. Lamellarin I (**55**), another polycyclic aromatic alkaloid isolated from tunicates in the genus *Didemnum*, has been found to effectively reverse multi-drug resistance in cancer cells by inhibiting P-glycoprotein mediated drug efflux.[54] The potency of lamellarin I as a MDR modulator was 9- to 16-fold higher than that of the benchmark compound verapamil.

Another promising recent development is the increase in reports of interesting 'lead' compounds isolated from cultures of marine microorganisms. Two examples illustrate the trend. A family of anti-fungal compounds named 15G256 γ (**56**), δ, and ε have been isolated from the mangrove derived fungus *Hypoxylon oceanicum* by scientists at Wyeth Ayerst.[55] These compounds have been patented and they are being evaluated for potential use against human fungal infections. Cyclomarin A (**57**), a heptapeptide obtained from cultures of a marine actinomycete, has been patented as an anti-inflammatory agent.[56] The use of molecular target assays and investigations of marine microorganism secondary metabolites represent important trends for the future in marine natural products drug discovery research.

5 Future Prospects

Nearly four decades of marine natural products chemistry investigations have demonstrated that organisms living in the world's oceans are an incredibly rich storehouse of novel low molecular weight chemical entities. The natural products chemical diversity pool present in marine algae and invertebrate extract libraries has virtually no overlap with the chemical diversity pools present in traditional synthetic libraries, combinatorial synthetic libraries, or terrestrial plant and microorganism natural product extract libraries. Many of the compounds already known exhibit potent biological activities that are relevant to treating human diseases. Therefore, marine natural products represent a unique and extremely valuable molecular resource for drug discovery programs using either traditional or HTS screening paradigms.

A large number of very promising compounds from the marine natural products chemical pool are currently in the pre-clinical and clinical stages of development, leading to the clear expectation that new pharmaceuticals from the sea will emerge for clinical use in the next decade. In order to keep the development pipeline full, researchers in the field will have to make marine extracts compatible with HTS discovery paradigms and they will have to continue to find creative ways to solve the supply issues. When the next generation of marine derived drugs do appear, it will represent the culmination of a new type of harvest from the bounty of the oceans. The ideal outcome of a marine natural product drug discovery effort is identification of a bioactive 'lead'

[54] A. R. Quesada, M. D. Garcia Gravalos and J. L. Fernandez Puentes, *Br. J. Cancer*, 1996, **74**, 677.
[55] D. Albaugh, G. Albert, P. Bradford, V. Cotter, J. Froyd, J. Gaughran, D. Kirsch, M. Lai, A. Rehnig, E. Sieverding and S. Silverman, *J. Antibiot.*, 1998, **51**, 317.
[56] W. H. Fenical, R. S. Jacobs and P. R. Jensen, US Patent 5444043, 1995.

structure that can be developed into a clinically useful synthetic entity. In this scenario, organisms in the ocean are yielding only knowledge about new pharmacophores rather than the more traditional harvests of biomass. This intellectual harvest is environmentally benign, it has the potential to greatly benefit all humans by leading to better treatments for serious diseases, and it can result in significant sustainable economic benefits to the countries where the relevant biological specimens were collected and where the advanced stages of development work have taken place.

Contamination and Pollution in the Marine Environment

STEPHEN J. DE MORA

1 An Overview of Marine Pollution

Both contamination and pollution entail the perturbation of the natural state of the environment by anthropogenic activity. The two terms are distinguishable by the severity of the effect: pollution induces the loss of potential resources.[1] Additionally, a clear cause–effect relationship must be established for a substance to be classified as a pollutant towards a particular organism.

In the marine environment, human-induced disturbances take many forms. Owing to source strengths and pathways, the greatest effects tend to be in the coastal zone. Waters and sediments in such regions bear the brunt of industrial and sewage discharges and are subject to dredging and spoil dumping. Agricultural runoff may contain pesticide residues and elevated nutrients, the latter of which may over-stimulate biological activity, producing eutrophication in the water column or anoxic conditions in underlying sediments. However, the deep sea has not escaped contamination. The most obvious manifestations comprise crude oil, petroleum products, and plastic pollutants, but also include the long-range transport of long-lived radionuclides from continental sources. Additionally, the aeolian transport of heavy metals has enhanced natural fluxes of some elements, particularly lead.

In the 1960s, the global problems related to the wide-scale application of pesticides first accentuated the possible difficulties associated with utilizing useful and apparently benign compounds. The lesson supposedly learned with DDT has been reinforced in the past decade with tributyltin (TBT) compounds in the marine environment. TBT, a 'modern contaminant', is the main active ingredient in a range of organotin-based marine anti-fouling paints owing to its appreciable toxicity towards a wide range of organisms. In effect, TBT is a broad-spectrum pesticide and herbicide, with all the ensuing problems one might expect, that had been unwittingly introduced and popularized in a pesticide-aware culture. The

[1] E. Goldberg, *Mar. Pollut. Bull.*, 1992, **25**, 45.

Issues in Environmental Science and Technology No. 13
Chemistry in the Marine Environment
© The Royal Society of Chemistry, 2000

widespread deleterious influence of TBT on non-target organisms, including those of commercial value, has resulted in the rapid regulation of TBT in many countries and international regulations to ban TBT are under consideration. However, the case history of TBT has demonstrated the difficulties associated with assuming stewardship of the global environment.[2]

The concept of marine pollution has undergone a continual evolution. Three key factors in such a re-evaluation pertain to the nature of the pollutants and the processes by which contamination is investigated and, more recently, controlled. Firstly, the understanding of what constitutes a threat to the marine environment, once perceived to entail essentially just chemical contaminants, has been broadened in scope. Thus, pollution by such agents as metals, pesticides, and oil continues to feature prominently in environmental impact studies. More recently, one could add organometals, radionuclides, and endocrine disrupters to the list of chemicals. However, other agents of anthropogenic change are now being investigated. Hence, non-chemical causes for concern include the heat associated with the emission of cooling waters from power plants, exotic biota from ship ballast waters and hulls, and enhanced sediment discharge as a result of accelerated soil erosion caused by deforestation and urban development.

Secondly, the research process to investigate contaminants has changed. Biogeochemical practices have ensured that understanding the behaviour of contaminants in the marine environment requires a multidisciplinary strategy. This takes into appreciation the role of hydrodynamics in pollutant transport processes at one extreme and elucidating biochemical responses at a sub-cellular level at the other. The development of more sensitive instrumentation has facilitated the creation of a better database of pollutant distributions. However, of greater significance is the realization that marked biological responses can be induced at extremely low contaminant concentrations. Sensitive assays now allow the investigation of sub-lethal effects of pollutants on cellular components and processes. Thus, several compounds have been shown to act as endocrine disrupters and, accordingly, ecotoxicology comprises a part of the spectrum of marine pollution studies.

Finally, recognition of the deleterious effects of marine pollution has led to a variety of control strategies. These strategies can be legal or technological in nature, but have usually been reactive responses to a pollution event. Thus, there is a notable ongoing effort to develop technologies to remediate polluted marine environments, particularly following oil spills. In the same vein, legislation has been introduced, both nationally and internationally, to limit known sources of contamination. The efforts by the International Maritime Organization (IMO) are notable in this regard with respect to limiting pollution from ships. However, the approach has been changing to become proactive in character. Essentially, the hard-learned lessons from DDT and TBT have instilled caution, for instance in the introduction of new antifouling agents in marine paints. This cautionary approach places greater reliance on risk assessment procedures and cost–benefit analyses. Present and future challenges rest with the need to identify the

[2] S. J. de Mora (ed.), *Tributyltin: Case Study of an Environmental Contaminant*, Cambridge University Press, Cambridge, 1996.

acceptable limits of pollution based on environmental, social, economic, and legal criteria. Marine pollution, perhaps once considered the domain of chemists and biologists, now impinges and relies on numerous other disciplines.

2 Selected Case Studies

There are numerous examples and instances of pollution in the marine environment and a comprehensive coverage would be beyond the scope of a single article. Only a few case studies are presented here. They are global in character and represent examples where public and scientific concern has been sufficient to provoke strategies to mitigate and/or prevent such pollution.

Oil Slicks

Major releases of oil have been caused by the grounding of tankers (*e.g. Torrey Canyon*, Southwest England, 1967; *Argo Merchant*, Nantucket Shoals, USA, 1976; *Amoco Cadiz*, Northwest France, 1978, *Exxon Valdiz*, Alaska, 1990) or by the accidental discharge from offshore platforms (*e.g. Chevron MP-41C*, Mississippi Delta, 1970; *Ixtox I*, Gulf of Mexico, 1979). Because oil spills receive considerable public attention and provoke substantial anxiety, oil pollution must be put into perspective. Crude oil has habitually been introduced into the marine environment from natural seeps at a rate of approximately 340×10^6 L yr^{-1}. Anthropogenic activity has recently augmented this supply by an order of magnitude; however, most of this additional oil has originated from relatively diffuse sources relating to municipal run-off and standard shipping operations. Exceptional episodes of pollution occurred in the Persian Gulf in 1991 (910×10^6 L) and due to the *Ixtox I* well in the Gulf of Mexico in 1979 (530×10^6 L). In contrast to such mishaps, the *Amoco Cadiz* discharged only 250×10^6 L of oil in 1978, accounting for the largest spill from a tanker. The cumulative pollution from tanker accidents on an annual basis matches that emanating from natural seepage. Nevertheless, the impacts can be severe when the subsequent slick impinges on coastal ecosystems.

Regardless of the source, the resultant oil slicks are essentially surface phenomena that are affected by several transportation and transformation processes.[3] With respect to transportation, the principal agent for the movement of slicks is the wind, but length scales are important. Whereas small (*i.e.* relative to the slick size) weather systems, such as thunderstorms, tend to disperse the slick, cyclonic systems can move the slick essentially intact. Advection of a slick is also affected by waves and currents. To a more limited extent, diffusion can also act to transport the oil.

Transformation of the oil involves phase changes and/or degradation. Several physical processes can invoke phase changes. Evaporation of the more volatile components is a significant loss mechanism, especially for light crude oil. Oil slicks spread as a buoyant lens under the influence of gravitational forces, but generally separate into distinctive thick and thin regions. Such pancake formation is due to the fractionation of the components within the oil mixture.

[3] S. Murray, in *Pollutant Transfer and Transport in the Sea*, ed. G. Kullenberg, CRC Press, Boca Raton, FL, 1982, vol. 2, p. 169.

Sedimentation can play a role in coastal waters when rough seas bring dispersed oil droplets into contact with suspended particulate material and the density of the resulting aggregate exceeds the specific density of seawater. Colloidal suspensions can consist of either water-in-oil or oil-in-water emulsions, which behave distinctly differently. Water-in-oil emulsification creates a thick, stable colloid that can persist at the surface for months. The volume of the slick increases and it aggregates into large lumps known as 'mousse', thereby acting to retard weathering. Conversely, oil-in-water emulsions comprise small droplets of oil in seawater. This aids dispersion and increases the surface area of the slick, which can accelerate weathering processes.

Chemical transformations of oil are evoked through photochemical oxidation and microbial biodegradation. Not only is the latter more important in nature, but strategies can be adopted to stimulate biological degradation, consequently termed bioremediation. All marine environments contain microorganisms capable of degrading crude oil. Furthermore, most of the molecules in crude oils are susceptible to microbial consumption. Oil contains little nitrogen or phosphorus, and as a result, microbial degradation of oil tends to be nutrient limited. Bioremediation often depends upon on the controlled and gradual delivery of these nutrients, while taking care to limit the concurrent stimulation of phytoplankton activity. Approaches that have been adopted are the utilization of slow-release fertilizers, oleophilic nutrients, and a urea-foam polymer fertilizer incorporating oil-degrading bacteria. Bioremediation techniques were applied successfully in the cleanup of Prince William Sound and the Gulf of Alaska following the *Exxon Valdez* accident. Alternative bioremediation procedures relying on the addition of exogenous bacteria have still to be proved. Similarly, successful bioremediation of floating oil spills has yet to be demonstrated.

Source apportionment of crude oil in seawater and monitoring the extent of weathering and biodegradation constitute important challenges in environmental analytical chemistry. As the concentration of individual compounds varies from one sample of crude oil to another, the relative amounts define a signature characteristic of the source. Compounds that degrade at the same rate stay at fixed relative amounts throughout the lifetime of an oil slick. Hence, a 'source ratio', which represents the concentration ratio for a pair of compounds exhibiting such behaviour, remains constant. Conversely, a 'weathering ratio' reflects the concentration ratio for two compounds that degrade at different rates and consequently this continually changes. Oil spill monitoring programmes conventionally determine four fractions:[4]

- Volatile hydrocarbons
- Alkanes
- Total petroleum hydrocarbons
- Polycyclic aromatic hydrocarbons (PAHs)

The volatile hydrocarbons, albeit comparatively toxic to marine organisms, evaporate relatively quickly and hence serve little purpose as diagnostic aids. The

[4] G. S. Douglas, A. E. Bence, R. C. Prince, S. J. McMillen and E. L. Butler, *Environ. Sci. Technol.*, 1996, **30**, 2332.

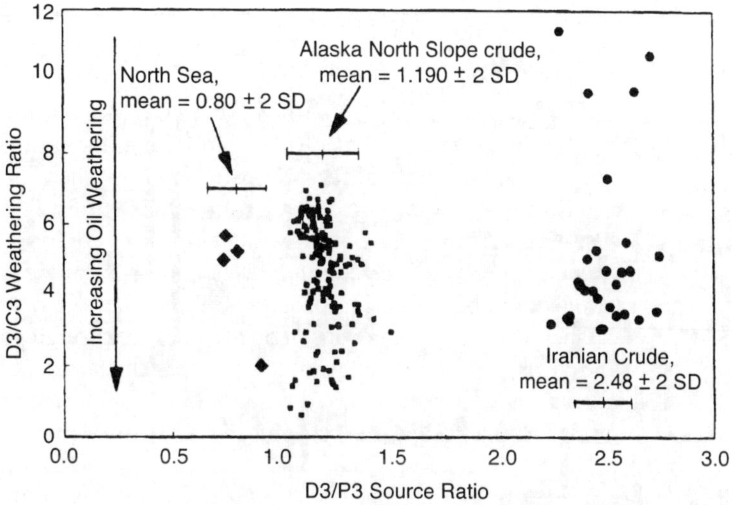

Figure 1 Plot of weathering ratio (C3-dibenzothiophenes: C3-chrysenes) *versus* source ratio (C3-dibenzothiophenes: C3-phenanthrenes) for fresh and degraded oil samples from three different crude oil spills (Reprinted with permission from *Environ. Sci. Technol.*, **30**, 2332. © 1996 American Chemical Society)

alkanes and total petroleum hydrocarbons make up the bulk of the crude oil. They can be used to some extent for source identification and monitoring weathering progress. The final fraction, the PAHs, comprises only about 2% of the total content of crude oil but includes compounds that are toxic. Moreover, these components exhibit marked disparities in weathering behaviour due to differences in water solubility, volatility, and susceptibility towards biodegradation. As demonstrated in Figure 1, both a D3/P3 source ratio (C3-dibenzothiophenes: C3-phenanthrenes) and a D3/P3 weathering ratio (C3-dibenzothiophenes: C3-chrysenes) have been defined from amongst such compounds that enable the extent of crude oil degradation to be estimated in the marine environment, as well as for subtidal sediments and soils.[4]

Litter and Debris

The increasing accumulation of litter along shorelines epitomizes a general deterioration of environmental quality of the marine environment. Such degradation extends to the high seas, being manifest as floating debris. The material originates not only from coastal sources, but also arises from the ancient custom of dumping garbage from ships. Thus, high concentrations of floating rubbish have been observed near fishing grounds and in shipping lanes. Drilling rigs and offshore production platforms have similarly acted as sources of contamination. Some degree of protection in recent years has accrued from both the London Dumping Convention (LDC) and the International Convention for the Prevention of Pollution from Ships (MARPOL) which outlaw such practices. However, the problem of seaborne litter remains global in extent and not even Antarctica has been left unaffected.[5]

The floating debris and beach litter consists of many different materials that, tending to be non-degradable, endure in the marine environment for many years.

[5] M. R. Gregory and P. G. Ryan, in *Marine Debris: Sources, Impacts and Solutions*, ed. J. M. Coe and D. B. Rogers, Springer, New York, 1996, p. 49.

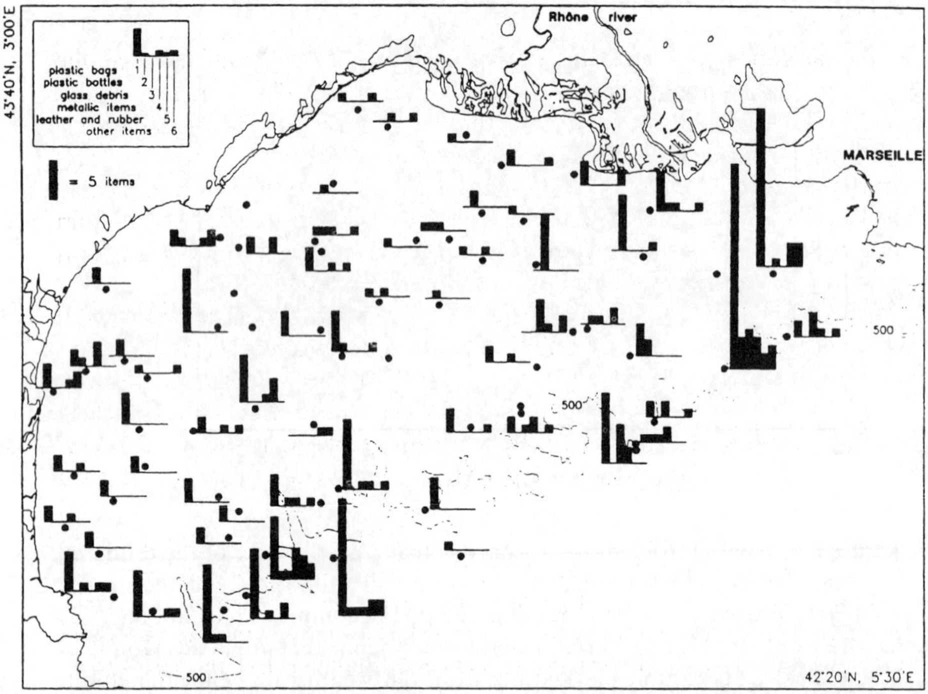

Figure 2 Quantities of debris per trawling tow (30 min) collected on the continental shelf and adjacent canyon of the Gulf of Lyons (Reprinted from *Mar. Pollut. Bull.*, **30**, 713. © 1995, with permission from Elsevier Science)

The most notorious are the plastics (*e.g.* bottles, sheets, fishing gear, packaging materials, and small pellets). Numerous other materials have been observed, such as glass bottles, tin cans, and lumber. This litter constitutes an aesthetic eyesore on beaches, but more importantly can be potentially lethal to marine organisms. Deleterious impacts on marine birds, turtles, and mammals result from entanglement and ingestion. While this floating material can have the apparently benign consequence of acting as a habitat for opportunistic colonizers, this allows the introduction of exotic species into new territories, with all the latent problems that such invasions can cause. Lost or discarded plastic fishing nets remain functional and can continue 'ghost fishing' for several years.

Perhaps more startling is the unseen pollution that has largely gone undetected on the sea floor. Bottom trawls in the northwest Mediterranean Sea found that litter was essentially ubiquitous in the region (see Figure 2).[6] The highest concentrations occurred in the vicinity of metropolitan areas, but canyons also tended to be sites of preferential accumulation. Again plastics dominated the material found, with up to 90% of the litter at a site near Marseilles comprising plastic bags. Plastic debris settling on soft and hard bottoms can smother benthos and limit gas exchange with pore waters. On a different note, traps and pots that go astray can continue to catch benthic animals.

[6] F. Galgani, S. Jaunet, A. Campillo, X. Guenenen and E. His, *Mar. Pollut. Bull.*, 1995, **30**, 713.

Tributyltin

Tributyltin (TBT) provides an interesting case study of a pollutant in the marine environment.[2] Because TBT compounds are extremely poisonous and exhibit broad-spectrum biocidal properties, they have been utilized as the active ingredient in marine anti-fouling paint formulations. Its potency and longevity ensures good fuel efficiencies for ship operations and guarantees a long lifetime between repainting. TBT-based paints have been used on boats of all sizes, from small yachts to supertankers, ensuring the global dispersion of TBT throughout the marine environment, from the coastal zone to the open ocean.

Notwithstanding such benefits, the extreme toxicity and environmental persistence has resulted in a wide range of deleterious biological effects on non-target organisms. TBT is lethal to some shellfish at concentrations as low as 0.02 μg TBT-Sn L^{-1}. Lower concentrations result in sub-lethal effects, such as poor growth rates and reduced recruitment leading to the decline of shellfisheries. The most obvious manifestations of TBT contamination have been shell deformation in Pacific oysters (*Crassostrea gigas*) and the development of imposex (*i.e.* the imposition of male sex organs on females) in marine gastropods. The latter effect, an example of TBT acting as an endocrine disrupter, has caused dramatic population decline of gastropods at locations throughout the world. TBT has been observed to accumulate in fish and various marine birds and mammals, with as-yet unknown consequences. Although it has not been shown to pose a public health risk, one recent study reported measurable butyltin concentrations in human liver.[7]

The economic consequences of the shellfisheries decline led to a rapid political response globally. The first publication suggesting TBT to be the causative agent appeared only in 1982,[8] but already the use of TBT-based paints has been banned in some countries, including New Zealand. Other nations have imposed partial restrictions, its use being permitted only on vessels >25 m in length or on those with aluminium hulls and outdrives. This has certainly had the effect of decreasing the TBT flux to the marine environment, as manifested in sedimentary TBT profiles. Oyster aquaculture in Arcachon Bay benefited immediately, with a notable decline in shell deformations and TBT-body burdens and the complete recovery of production within two years (see Figure 3).[9] Comparable improvements in oyster conditions have been reported for Great Britain and Australia. Similarly, there have been many reported instances of restoration of gastropod populations at previously impacted locations. However, large ships continue to act as a source of TBT to the marine environment. It should be of concern that imposexed gastropods have been observed at sites (*e.g.* North Sea and Strait of Malacca) where the source of TBT can only be attributed to shipping.

TBT exists in solution as a large univalent cation and forms a neutral complex with Cl^- or OH^-. It is extremely surface active and so is readily adsorbed onto suspended particulate material. Such adsorption and deposition to the sediments limits its lifetime in the water column. Degradation, via photochemical reactions

[7] K. Kannan and J. Falandysz, *Mar. Pollut. Bull.*, 1997, **34**, 203.

[8] C. Alzieu, M. Heral, Y. Thibaud, M. J. Dardignac and M. Feuillet, *Rev. Trav. Inst. Marit.*, 1982, **45**, 100.

[9] C. Alzieu, *Mar. Environ. Res.*, 1991, **32**, 7.

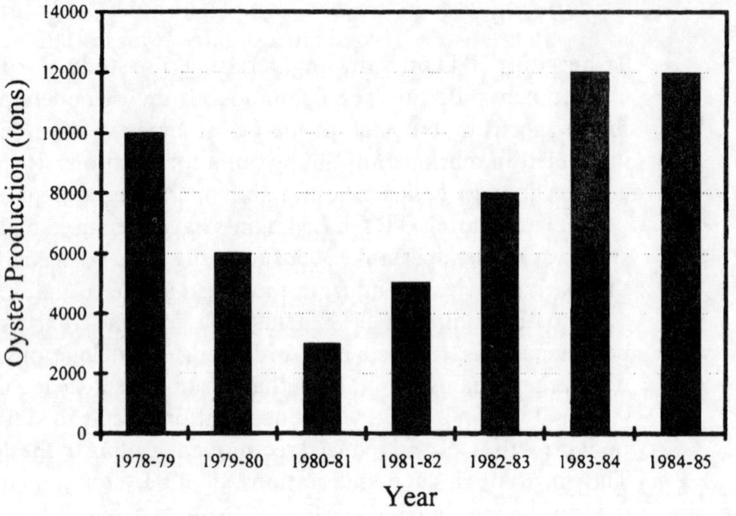

Figure 3 Annual oyster (*Crassostrea gigas*) production in Arcachon Bay, 1978–85; restrictions on TBT use were first applied in January 1982 (Data taken from Alzieu[9])

or microbially mediated pathways, obeys first-order kinetics. Several marine organisms, as diverse as phytoplankton to starfish, debutylate TBT. Stepwise debutylation produces di- and mono-butyltin moieties, which are much less toxic in the marine environment than is TBT. As degradation lifetimes in the water column are of the order of days to weeks, degradation is slow relative to sedimentation. Thus, TBT accumulates in coastal sediments where degradation rates are much slower, with the half-life being of the order of years.[10] Furthermore, concentrations are highest in those areas, such as marinas and harbours, which are most likely to undergo dredging. The intrinsic toxicity of TBT, its persistency in the sediments, and its periodic remobilization by anthropogenic activity are likely to retard the long-term recovery of the marine ecosystem. One recent study[11] has estimated that the residence time of TBT in oligotrophic waters may also be of the order of years. This may, in part, help to explain the measurable quantities of TBT in squid and marine mammals collected in the open sea.

Recapitulating, the unrestricted use of TBT has ended in many parts of the world but significant challenges remain. For the most part, the coastal tropical ecosystems remain unprotected and the sensitivity of its indigenous organisms is relatively poorly evaluated. TBT endures in sediments globally, with concentrations usually greatest in environments most likely to be perturbed. The widespread introduction of TBT into seawater continues from vessels not subject to legislation. Organisms in regions hitherto considered to be remote now manifest TBT contamination and effects. Such observations imply that further restrictions on the use of organotin-based paints are required. Indeed, the International Maritime Organization has undertaken to draft a global, legally binding convention that would prohibit the application of TBT-based paints after January 1, 2003, and ban the presence of TBT on ship hulls as from January 1, 2008. Of course, the prescribed dates will provoke considerable debate. However,

[10] C. Stewart and S. J. de Mora, *Environ. Technol.*, 1990, **11**, 565.
[11] P. Michel and B. Averty, *Environ. Sci. Technol.*, 1999, **33**, 2524.

the paramount lesson learned from TBT should be that potential replacement compounds must be properly investigated prior to their introduction in order to avoid another global pollution experiment.

3 Mitigation of Marine Pollution

As indicated above, national policies and international conventions have been invoked to curb known, and usually obvious, sources of marine pollution. National legislation is used to control coastal discharges of contaminants. A major problem remains owing to the inadequate treatment of sewage prior to emission from land-based sources. The most important deleterious effects in this case are with respect to microbial water quality. This can have a direct influence on bathing criteria and result in beach closures during contamination episodes. An additional problem from land-based sources pertains to transboundary effects, whereby pollution may inadvertently be exported from one country to another.

A series of international conventions have been negotiated, especially to control pollution from ship-based sources. Numerous restrictions apply to the release of rubbish and oil discharges. Similarly, the role of ships as a source of biological contaminants has become appreciated. Thus, adequate control of sewage discharges and ballast water is increasingly advocated. The environmental threat posed by antifoulants has become widely recognized owing to TBT-based paints. As noted above, such material should become prohibited in the near future. Replacement compounds will need to be shown to be more environmentally friendly. Although this is an obvious sentiment, the criteria to judge (or rank) such biocidal compounds remain contentious. The desired attributes are usually considered to be a high degradation rate, leading to non-toxic products and a low bioaccumulation potential (possibly manifest by a low water–octanol partition coefficient). A negative consequence of the use of biocides other than TBT may be the enhanced capacity for shipping to act as a transportation vector for the invasion of exotic species into new territory.

Economic Controls

Although economic forces have often been viewed to be an agent of environmental degradation, the perception and role of market forces in preventing and mitigating pollution is changing. Garrod and Whitmarsh[12] have appraised governmental and economic methods of controlling pollution in the marine environment. They suggested that economic controls to pollution control have been gaining favour over 'command and control' strategies, but the role of governmental action is not likely to become redundant.

As background, Garrod and Whitmarsh[12] describe market failure with respect to protecting the marine environment. Particular difficulties arise because of the diverse and perhaps conflicting exploitation of the sea and its resources. Thus, pollutants may be released into the marine environment by one sector or industry

[12] B. Garrod and D. Whitmarsh, *Mar. Pollut. Bull.*, 1995, **30**, 365.

but the environmental costs are born by society and/or other users. Property rights are usually so poorly defined that the polluter cannot readily identify the party to whom compensation should be paid.

Government intervention for environmental protection is not without criticism. 'Standard setting' solutions can have a proclivity towards over-regulation and may be unable to distribute appropriately or adequately the associated costs and compensations. Some policies may favour specific sectors of industry, to the extent where regulations promote the industry at the expense of customers, society, or other industrial sectors. As an alternative, governments can promote economic approaches. Taxes and charges may be levied in an attempt to ensure that the true costs of production are borne by the industry. The 'polluter pays' adage of the 1980s thereby transfers the financial burden of environmental damage from society to the producer, with the incentive that reducing pollution activities may realise cost benefits. An alternative tactic is to impose a system of tradable pollution permits. Again, the incentive then rests with the producer to minimize environmental damage and associated costs.

Corporate environmentalism is an evolving concept for environmental protection. In this case, business takes a pro-active stance independent of regulatory authorities. This can be in recognition of social responsibilities, but is more successful when compelled by competition in the market place. Thus, a firm can conscientiously target environmentally aware consumers (through marketing environmentally friendly products or processes) or can be better placed for financial support from ethical investment funding bodies.

Despite the attractions of economic forces driving environmental protection, some cautions and failures have been noted.[12] Firstly, the export of hazardous waste to countries where costs for treatment are lower enhances environmental risks during transport and has the potential for transboundary export in the event of pollution. At the same time, the loss of raw material may deprive the home market of an adequate supply of feedstock for the home-based industry. Secondly, there is considerable scepticism that self-regulation of TBT-based antifoulants could be achieved in a timely manner by the shipping industry. This is an instance where the cost benefits to one industry are born by another commercial sector, notably aquaculture. Thus, protection of the marine environment is likely to be aided by economic factors but the role of government, via taxation and standard setting, is not likely to be usurped. Public education and, in turn, pressure, can promote and support corporate environmentalism.

Bioremediation

Legislation and economic factors may aim to prevent marine pollution. Nevertheless, contamination is inevitable and technological solutions to mitigating the impacts have been developed. This is especially the case for oil pollution, which inevitably receives considerable press attention. Accidental oil spills at sea do occur and frequently impact shoreline environments. Petroleum pollutants can be removed by microbial degradation. Although bacteria and fungi capable of degrading many oil components exist in the marine environment, natural rates of hydrocarbon biodegradation are usually limited by abiotic environmental factors. Numerous strategies to accelerate oil biodegradation rates have been developed over the last 20 years and *in situ* bioremediation has become an established oil spill countermeasure.[13]

Bioremediation refers to the addition of substances or modification of habitat at contaminated sites to accelerate biodegradation. Two approaches have been used. In bioaugmentation, oil-degrading bacteria are introduced to supplement the existing microbial population. However, oil-degrading microflora naturally increase in numbers following exposure to oil. Moreover, laboratory and field trials have failed to provide convincing evidence of consistent success. The cost-effective application of such technology remains controversial and not well justified. In contrast, biostimulation involves the addition of nutrients or growth-enhancing co-substrates and/or improvements in habitat quality to enhance the growth of indigenous oil-degrading bacteria. Several different strategies have been tested.

Firstly, given that nutrient availability often limits microbial activity, fertilization with nitrogen and phosphorus has been used. To prevent rapid dilution and to maintain a sufficient concentration of nutrients to support the maximal oil biodegradation rates, they generally are incorporated into oleophilic nutrient formulations or microemulsions, which are retained in interfacial regions (*e.g.* air–sea interface or on the surfaces of sediments in beaches). Its efficacy during actual response operations has been demonstrated on cobble beaches contaminated by the *Exxon Valdez* spill in Alaska.[14] Secondly, diverse means to oxygenate sedimentary environments have been attempted because anoxic conditions dramatically limit microbial oil degradation rates. Deeper penetration of oxygen and nutrient supplements can be achieved with tilling and raking. Alternatively, chemical oxidants, such as hydrogen, calcium, and magnesium peroxides, can alleviate oxygen deficiency within sediments. Transplantation may aerate the rhizosphere and serve as a means to stimulate aerobic oil biodegradation. The introduced plants also may take up oil and release exudates and enzymes that further stimulate microbial activity. This technique, known as phytoremediation, has potential application in delicate and sensitive salt marsh environments that are the most difficult to clean. Finally, methods to increase the surface area of the oil–water interface have been applied, this being where microbial oil degradation principally occurs. Thus, chemical dispersants, surface agents such as powdered peat, and fertilizers supplemented with biosurfactants have all been used as bioremediation agents.

Recommended for use following the physical removal of bulk oil, bioremediation has an operational advantage in that it breaks down and/or removes the residual contaminants in place. This technology is relatively cost-effective, not requiring a large number of personnel or highly specialized equipment. Laboratory experiments and field trials have demonstrated the feasibility and success of bioremediation to enhance bacterial degradation of oil on cobble, sand beach, and salt marsh environments. Termination of treatment should be implemented when: (1) it is no longer effective; (2) the oil has degraded to acceptable biologically benign concentrations; or (3) toxicity due to the treatment is increasing.

[13] K. Lee and S. J. de Mora, *Environ. Technol.*, 1999, **20**, 783.

[14] R. C. Prince, *Crit. Rev. Microbiol.*, 1993, **19**, 217.

4 Summary

Marine pollution takes many forms. A few case examples have been described here with the objective to portray the diversity of contamination. Many types, locations, and impacts can be contemplated, but some characteristics are universal. On a positive note, there are many mechanisms for preventing pollution and mitigating long-term adverse effects. Thus, national legislation and international conventions provide considerable protection from both land- and sea-based sources of pollution. Economic forces can be used to control pollution, either via governmental intervention in the form of taxation or through corporate environmentalism. In the inevitable consequence of marine pollution events, bioremediation strategies have successfully aided shoreline recovery from oil spills. This field is still evolving, with the major challenge of cleaning spills at sea still remaining.

Subject Index